NUMBER ONE:

The Texas Engineering Experiment Station Monograph Series
ANDREW R. MCFARLAND, *Editor-in-chief*
Texas A&M University

Equilibrium Properties of Fluids and Fluid Mixtures

Equilibrium Properties of Fluids and Fluid Mixtures

By ALEKSANDER KREGLEWSKI

PUBLISHED FOR THE
TEXAS ENGINEERING EXPERIMENT STATION,
THE TEXAS A&M UNIVERSITY SYSTEM,
BY
TEXAS A&M UNIVERSITY PRESS
COLLEGE STATION

Kreglewski, Aleksander, 1927–
    Equilibrium properties of fluids and fluid mixtures.

    (The Texas Engineering Experiment Station monograph
series; no. 1)
    Bibliography: p.
    Includes index.
    1. Liquid-liquid equilibrium. 2. Mixtures.
I. Title. II. Texas Engineering Experiment Station. III. Series.
QD503.K73   1984   530.4′2   83-40493
ISBN 0-89096-183-2

Manufactured in the United States of America
FIRST EDITION

To my parents

In Memoriam

Aleksander Kreglewski passed away on February 3, 1984, at the age of 56, having seen only the galley proofs of this book, his major work. The tragedy of this situation was, in many ways, consistent with his entire life.

Alek was Polish by birth and retained a love for his homeland along with that for his adopted country, the United States. He received his education in Poland and as a teenager witnessed and participated in the Second World War. While he was attempting to disarm a land mine, it exploded, and he incurred a serious hearing loss which became almost total in the years immediately preceding his death. His accomplishments in life were made even more remarkable by the presence of that handicap.

He received his Doctor of Chemistry degree from the Polish Academy of Sciences in 1956 and remained there as a research chemist until 1958. He then spent two years at Carnegie Institute in a postdoctoral capacity and subsequently returned to Poland as Head of the Phase Equilibria Laboratory in the Warsaw Institute of Physical Chemistry. In 1966, he emigrated to the United States (becoming a citizen in 1972) and joined the staff of Texas A&M University. During an eighteen-month period in 1966–68, he was a visitor at Ohio State and, upon returning to Texas A&M, completed his career as a member of the Thermodynamics Research Center staff. During his scientifically active years, he published 55 papers and contributed to several collective works. His efforts received wide recognition for excellence and gained him the respect of his peers. This book represented an opportunity for him to record those works he considered to be most important and to share his insight with the scientific and engineering community.

On a personal basis, Alek was an interesting intellect and a man of gentlemanly manner. He was a devoted and loving father to his daughter, Agnesieke, and he was a kind and loyal friend to those who knew him. Truly, we are all diminished both personally and professionally by his passing.

Kenneth R. Hall
Andrew R. McFarland
College Station, Texas
April 4, 1984

Contents

List of Symbols

Subscripts

$i, j, k, \ldots$	running suffixes for different species (or 1, 2, $\ldots$, also used for individual molecules)
$\alpha, \beta, \gamma, \ldots$	running suffixes for different molecules (of the same or different species)
m	molar or per mole of an m-component mixture; also an average value of a molecular property of a mixture (the excess functions are also molar, but the subscript is deleted)
σ	property along the saturation curve
o, oo	property of a reference substance
x	see f_x, h_x; also pseudocritical constants

Superscripts

none	configurational property
r	residual property
E	excess property
c	gas-liquid critical locus
cs	critical solution temperature (liquid-liquid or gas-gas); e.g., T^{ucs} (upper), T^{lcs} (lower)
b	normal boiling point
v	vaporization
∞	infinite dilution
$-$	bar over a symbol: extremum value, e.g., $\bar{x}_i$, the azeotropic composition
$\ominus$	the values at $T/T^c = 0.6$ or any "standard" quantity
$*$	reduced quantity, Eqs. (4.22) and (4.23)
$'\ ''$	primes, double primes indicate phases in the order of decreasing molar density
L, μ, Q	London, dipole, quadrupole energy contributions

Symbols

N_0	Avogadro number, 6.02205×10^{23} molecules $\cdot$ mol^{-1}
k	Boltzmann constant, 1.38066×10^{-23} J $\cdot$ K^{-1}
R	$= N_0 k$, gas constant, 8.31441 J $\cdot$ mol^{-1}K^{-1}
h	Planck constant, 6.62618×10^{-34} J $\cdot$ s
M	molar mass
P	pressure
V	volume
T	thermodynamic temperature (Kelvins)
Z	compressibility factor
N	number of molecules (total)
n	amount in moles
ρ	density, $1/V$
ρ_n	number density, N/V
A	Helmholtz free energy
G	Gibbs free energy
U	internal energy, Eq. (1.12a)
H	enthalpy (heat content)
S	entropy
f_i	fugacity
μ_i	chemical potential
h_i, v_i	partial molar enthalpy, volume
H_{ij}	Henry's constant
$\beta, \beta(T)$	second virial coefficient
$\beta^{(3)}$	third virial coefficient
K_i	vaporization equilibrium constant
α_{ij}	volatility ratio, Eq. (8.2)
z	coordination number
z_{ij}	frequency of collisions
m_c	number of chain links (also number of carbon atoms)
x_i	mole fraction
q_i	contact fraction, Eq. (5.20)
φ_i	g-fraction, Eq. (5.17)
ϕ_i	site fraction, Eq. (5.28)
$u(r)$	intermolecular energy; $u(r)/k$ in Kelvins
$\bar{u}$	minimum value of $u(r)$
σ	collision diameter

η, δ	parameters of noncentral energy
f_x	$\bar{u}_m/\bar{u}_{oo}$
h_x	$(\sigma_m/\sigma_{oo})^3$
p_i	polarizability (mean)
μ_i	dipole moment (scalar)
Q_i	quadrupole moment (scalar)
ω	acentric factor
E	electronegativity
Ω	phase integral (partition function)
Ψ	wave function, Eq. (9.4); product of partition functions, Eq. (1.6)
Q	configurational integral
$\mathcal{U}$	configurational (potential) energy of an assembly of molecules
$d\mathbf{r}$	element of volume, $dxdydz$ or $r^2 \sin \Theta \, dr \, d\Theta \, d\varphi$

Units

The following practical SI units are used:

Pressure	1 bar $= 10^5$ Pa $= 14.50377$ lb(wt) $\cdot$ in^{-2}
Volume	1 liter $= 10^3$ cm$^3 = 3.531467 \times 10^{-2}$ ft^3
	$= 0.2641721$ gal (U.S.)
Temperature	Kelvin (K); $T/K = 273.150 + t/°C$
	where $t/°C = (t/°F - 32)/1.8$
Energy	1 kJ $= 10^3$ J $= 0.2390057 \times 10^3$ cal
	$= 0.9478172$ BTU
Molar mass	1 mol $= 1$ g-mol $= 2.20462 \times 10^{-3}$ lb-mol

Preface

The physical properties of mixtures of fluids have been at the center of the research interests of many physical chemists and chemical engineers for more than a century. One of the important reasons was and continues to be the need for efficient and economical separation and purification of the components of raw materials, such as petroleum or coal tar, and also of synthetic products. Another reason is purely scientific: namely, fascination with the elusive properties of fluids and mixtures of fluids.

The theory of fluids has made great progress in recent years. These papers are scattered in journals of theoretical physics, however, and the equations seem hopelessly complex for a chemical engineer interested in simple practical solutions. Consequently, the gap between pure science and engineering in this field is growing.

The results obtained in recent years show that the accuracy of either smoothing or predicting phase equilibria in mixtures depends mostly on that of the equations of state, the extent of our knowledge of the intermolecular forces, and their temperature dependence. Equations of state of hard bodies were developed, accounting properly for the repulsion between molecules, as well as the perturbation theories and molecular dynamics simulations taking care of the intermolecular attraction. These accurate equations of state and fast computers allow solution of the Gibbs equilibrium conditions directly. Therefore, we shall return in this book to the original idea of van der Waals (1890, 1908) and his coworkers that the properties of fluids are determined to the same extent by repulsion and by attraction forces. Accordingly, this book could bear a subtitle, *Applications of the Augmented van der Waals Theory of Fluids*, to honor the name of the pioneering scientist.

Besides equations of state, the theory of intermolecular energy $u(r)$ will be considered to a far greater extent than it is in any of the existing monographs on mixtures. The considerations will be limited to fluids and mixtures of non-electrolytes with or without weak hydrogen bridges (bonds), but the theoretical foundations allow an extension to such substances as water, alcohols, and organic acids, the latter only in mixtures with inert solvents. The radial distribution functions $g(r)$, which are coupled with $u(r)$ and whose computation is very difficult, are replaced by functions of the reduced density ρ^* and reduced temperature T^*, $F(\rho^*, T^*)$, based on molecular dynamics or the PVT properties of a reference fluid (argon). The functions $F(\rho^*, T^*)$ for polar or quadrupolar fluids will be further simplified so that the dependence on ρ^* will be the same as for argon. All these simplifications seem to be necessary in routine computations. The application of the above concepts to the calculation of complete phase diagrams of mixtures at high pressures, including the critical locus curves, is the main object of my book.

The theories of mixtures that were published before the development of recent equations of state are now briefly reviewed. Hildebrand and coworkers developed the theory of regular solutions (also known as the Scatchard-Hildebrand theory), whose extensions and applications are summarized in three monographs by Hildebrand and Scott (1950, 1962) and by Hildebrand, Prausnitz, and Scott (1970). At the same time, Guggenheim (1952) applied the lattice model to mixtures;[†] in turn, Flory (1941, 1953) and Huggins (1941, 1958) published the first successful theory of polymer solutions based on the lattice model.

In the decade of the 1950s new approaches were proposed. The first, based on the cell model of the liquid state, was developed by Prigogine and Garikian and more recently by Prausnitz and his coworkers. The second approach, based on the principle of corresponding states, evolved from the theory of conformal mixtures developed by Longuet-Higgins (1951). Attempts to combine the advantages of the two approaches were made by Prigogine and Bellemans (the average potential model) and by Scott (the two-fluid model). The detailed accounts of these efforts are well summarized in the four books by Prigogine, Bellemans, and Mathot (1957), Rowlinson (1959), Prausnitz (1969), and King (1969), as well as in the review articles by Barker (1955), Brown (1965), Bellemans et al. (1967), Leland and Chapplear (1968), and R. L. Scott and Fenby (1969). During the same period, a surprisingly successful theory based on an abstract concept of

---

[†] For comments on the lattice model, see Gokcen and Chang (1971).

equilibrium between gaslike and solidlike degrees of freedom in a liquid was proposed by Eyring, Ree, and coworkers (1958). It has been further extended to mixtures with similar success (Liang et al., 1964; Ma and Eyring, 1965).

The theories based on the principle of corresponding states are the only ones that may be applied without an equation of state to the calculation of phase equilibria at high pressures. They will be discussed later. The validity of the others is questionable even when the vapor pressures of the components are very low, because these theories did not consider repulsion. Although the repulsion and the attraction terms in the relation for the compressibility factor Z are equal at $P = 0$, they differ in the residual Helmholtz energy A_m^r of the mixture. The differences between the repulsion and the attraction terms in $Z(T, \rho)$ and $A^r(T, \rho)$ are essential in phase equilibria of fluids and mixtures. All other aspects in which the theories differ among themselves are of minor importance. For these reasons the theories of mixtures usually lead to wrong values of the excess functions G^E and V^E. The contribution of the repulsion term to the residual internal energy U^r is negligible, and so some of the theories yield good values of H^E. Nevertheless, there are concepts in these theories that become useful when incorporated into an accurate equation of state. They will be discussed in Chaps. 4 and 6.

The development of the equation of state for hard spheres seemed of little value for the theory of long-chain molecules, and the latter was treated independently. Flory and coworkers (1964, 1965, 1970) published a successful theory of mixtures of chain molecules based on a modified van der Waals equation of state. Huggins (1970, 1971) developed an accurate theory of polymer solutions based on detailed statistics of the surfaces and interaction energies of the contacting segments of the mixture. When the molecules of both components (i, j) are small, it is proper to evaluate the average minimum value of $u(r)$: namely, $\bar{u}_m$ of the mixture from $\bar{u}_{ii}$, $\bar{u}_{jj}$, and $\bar{u}_{ij}$ weighed by means of the mole fractions. On the other hand, when j or both i and j are large r-mers, the surface fractions or the site fractions are used, giving greater weight to the larger molecule. Various approximations for weighing the contributions to $\bar{u}_m$ of mixtures of intermediate-size molecules are considered in Chap. 5, but not polymer mixtures, as the scope of this book is limited to more or less volatile fluids.

In order to complete this brief survey of the earlier literature, it is appropriate to mention the monographs by Malesinski (1965), Hala, Pick, and Vilim (1958), and van Ness (1967), as well as the textbooks by Haase (1956) and Prigogine and Defay (1954) on the classical thermodynamics of mixtures, which devote little or no attention at all to the molecular behavior of the mixtures.

All the thermodynamic equilibrium and stability conditions needed in the calculations of phase equilibria in pure fluids and mixtures (except those of a tricritical point) were derived by Gibbs (1948) in 1870–80, the same decade as the van der Waals equation. In order to apply these conditions in practice, it is necessary to distinguish a perfect gas from a real fluid. The differences between the two, expressed by the residual functions, can be evaluated when the molecular properties of the fluid and the equation of state are known. The basic relations for the residual functions are given in Chap. 1, and those resulting from the theory of intermolecular forces and the recent equations of state are considered in Chaps. 4 and 6.

Most students of chemistry and chemical engineering wonder at the multitude of relations in their courses in thermodynamics, and later they wonder again that still other relations are used in practice. We need a few basic relations to calculate the phase diagrams of mixtures and the enthalpies and entropies. The relations for multicomponent systems (for example, for the excess functions) are derived; other functions obtained from thermodynamics, theories of intermolecular forces, and the theory of intermolecular repulsion are quoted without derivation.

Finally, let us note that the bibliography at the end of this book does not reflect the tremendous amount of work done by the experimentalists. The data for pure fluids and those for mixtures are equally important. Each set of constants of an equation results from the data in ten, twenty, or more papers that are not cited here; however, they are cited in the papers where the constants were determined. Also, the sources quoted here for mixtures are only a few among the hundreds of papers. This does not mean that the others are less important but simply that the author was able to interpret only some of them at the present time.

The foundations of science are empirical. Theories must be improved or discarded, whereas the results of careful experiments are always good.

Acknowledgments

I offer my thanks to Dr. Bruno J. Zwolinski, former Director of the Thermodynamics Research Center (TRC), for encouraging this part-time research on mixtures of fluids and to the Texas Engineering Experiment Station for partial financial support.

Most of my experimental research was done with a group of students and coworkers before 1966, at the Institute of Physical Chemistry in Warsaw, Poland. A collaboration with Prof. Webster B. Kay in 1968–69 was very fruitful, and I will always recall with nostalgia those eighteen months at Ohio State University.

An extensive evaluation of various theories of mixtures and equations of state was perhaps the most important task. This I did in the years 1970–77, with the assistance of Dr. Stephen S. Chen in 1975–77. Our work progressed rapidly, but unfortunately, Stephen died in September 1977 at the age of 39. I have continued this work with much assistance from Mrs. Carol Chen in computer programming. The assistance of Dr. R. C. Wilhoit in programming during 1973 and 1981 is also acknowledged.

Thanks to the emphasis put on the studies of mixtures by Dr. Kenneth R. Hall, the Director of TRC since 1979, and a large grant from the National Science Foundation, I have been able to expand this work greatly and to test the usefulness of basic concepts by the calculation of complete phase diagrams for many binary systems.

In a nice gesture, Mrs. Ann-Lin Risinger prepared all the graphs, with great care and concern for every detail. Ms. Renèe C. Ney typed all the references. Finally, I wish to thank the reviewers for friendly and constructive comments.

College Station
July 1983

Equilibrium Properties of Fluids and Fluid Mixtures

CHAPTER 1

Thermodynamic Residual and Configurational Functions

It is well known that a liquid will evaporate during an isothermal expansion of the space in which it is confined until the dew point is reached. Upon further expansion, the properties of the unsaturated vapor or gas, when its density or the pressure is low enough, may approach those of a perfect gas. However, even at the density $\rho = 0$ not all of the properties of a real fluid become equal to those of the *perfect gas*. One of the properties that become identical for a real gas and the perfect gas at this density is the pressure

$$P(T, V, n_i) = (RT/V) \sum_i n_i \tag{1.1}$$

This limiting behavior of real fluids was confirmed by precise measurements of the compressibility (compression) factors Z, shown schematically in Fig. 1.1. The straight lines can be expressed by the relation

$$Z = PV_m/RT = 1 + \beta(T)\rho_m \tag{1.2a}$$

or

$$Z = 1 + \beta(T)P/RT \qquad (\rho_m \to 0) \tag{1.2b}$$

where ρ_m is the molar density, $1/V_m$. The function $\beta(T)$ depends on the chemical nature of the gas and on the temperature only. It is called the *second virial coefficient*. As $\rho \to 0$, $(\partial Z/\partial \rho)_T = \beta(T)$, and it is zero at all temperatures for a perfect gas only. Let us note that at $P \to 0$ the equations above are not necessarily valid, because the state at this pressure may be that of a nonvolatile liquid or the crystalline state. It is then more proper to operate with densities than with pressures.

As shown by Kirkwood (1935) and independently by Mayer and Mayer (1940) and de Boer (1949), the second virial coefficient is

$$\beta_{ij} = -2\pi N_0 \int_0^\infty (e^{-u_{ij}(r)/kT} - 1) r^2 \, dr \tag{1.3}$$

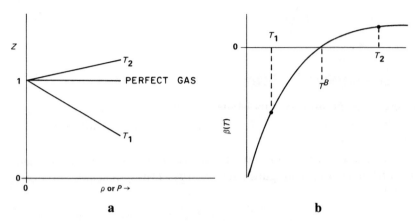

Fig. 1.1.   (a) The compressibility factor Z of a real gas at two temperatures, T_1 and T_2, and low densities. (b) The second virial coefficient of a real gas as the function of temperature. T^B, at which β equals zero, is called the *Boyle temperature*.

The intermolecular energy $u_{ij}(r)$ between the molecules of species i and j depends on the geometric properties (size and shape) of the molecules and the intermolecular forces of attraction. Thus, the perfect gas is a hypothetical gas of point-molecules ($\sigma = 0$) between which there is no attraction, $u(r) = 0$, at all distances r.

The total of the thermodynamic functions of a substance is the sum of the *molecular functions* and the *configurational functions*; for example,

$$A^{\text{total}} = A + A^{\text{mol}}; \quad G^{\text{total}} = A^{\text{total}} + PV;$$

also

$$G = A + PV;$$

hence

$$G^{\text{mol}} = A^{\text{mol}} \tag{1.4}$$

The total functions are derived from the phase integral (partition function) Ω, which is related to the mass and to the intra- and intermolecular energies of N_i molecules of species i. The fundamental relation for the Helmholtz energy[†] is

[†]The basic relations of thermodynamics and statistical mechanics such as (1.6), (1.8), and (1.12) are quoted here without derivation. The reader is referred to the textbooks on the subject.

$$A^{\text{total}} = -kT \ln \Omega \tag{1.5}$$

where

$$\Omega = \prod_i (2\pi m_i kT/h^2)^{3N_i/2} (\Psi_i)^{N_i} Q \tag{1.6}$$

where m_i is the mass of a molecule, h is the Planck constant, and Ψ_i is the product of the rotational and vibrational partition functions of a molecule of species i.

The integral Q is the *configurational integral*, as it depends on the intermolecular (configurational) energy $\mathcal{U}$ of N_i molecules in a volume V:

$$Q = \frac{1}{\prod_i N_i!} \int \cdot \underset{V}{\cdots} \cdot \int \exp(-\mathcal{U}/kT) \, d\,r_1, \ldots, d\,r_N \tag{1.7}$$

Accordingly, the configurational Helmholtz energy is

$$A = -kT \ln Q \tag{1.8}$$

and $A^{\text{mol}} = -kT \ln(\Omega/Q)$.

As shown by Rowlinson (1969) and by Rowlinson and Watson (1969), these relations allow us to derive most directly the basic relations for the thermodynamic functions of the perfect gas and a real fluid. Since we shall deal with nonreacting systems only, we are not interested in the molecular functions.

The configurational properties of the perfect gas are obtained from Eq. (1.7) for $\mathcal{U} = 0$, by using Stirling's formula[†] for $\ln N_i!$:

$$\ln Q^{\text{pg}} = N \ln V - \sum_i N_i(\ln N_i - 1) \tag{1.9}$$

Since $P = -(\partial A/\partial V)_T$ and $N = \Sigma_i \, N_i$, we obtain Eq. (1.1); that is, $P^{\text{pg}} = nRT/V$, where $n = \Sigma_i \, n_i$.

The author likes particularly the textbooks by Reif (1965), Hill (1960), and the two monographs by Rowlinson (1963, 1969). Münster's *Classical Thermodynamics* (1970) is a difficult textbook, but certainly the best for advanced courses. A good introductory description of states of matter is provided by Brostow (1979). The book also contains interesting considerations of the foundations of thermodynamics. The theory of partial molar quantities has just been re-derived and generalized by Reis (1982). It shows clearly the differences between the partial molar quantities $y_i(T, P)$ and $y_i(T, V)$.

[†]Stirling's formula for large N is: $\ln N! = N(\ln N - 1)$.

Further, since $N_i/N = x_i$,

$$A^{pg}(T, V, n_i)/nRT = \ln(N/V) - 1 + \sum_i x_i \ln x_i \qquad (1.10)$$

and

$$G^{pg}(T, P, n_i)/nRT = \ln(P/kT) + \sum_i x_i \ln x_i \qquad (1.11)$$

By using the Gibbs-Helmholtz relation

$$\frac{U}{RT} = -T\left[\frac{\partial(A/RT)}{\partial T}\right]_v \qquad (1.12a)$$

or

$$\frac{H}{RT} = -T\left[\frac{\partial(G/RT)}{\partial T}\right]_P \qquad (1.12b)$$

we obtain

$$U^{pg}(T, V, n_i)/nRT = 0, \qquad (1.13)$$

$$H^{pg}(T, P, n_i)/nRT = 1 \qquad (1.14)$$

and for the entropy

$$S^{pg}(T, V, n_i)/nR = -A^{pg}(T, V, n_i)/nRT \qquad (1.15)$$

According to Gibbs, the chemical potential of the *i*th component of a mixture is defined as

$$\mu_i = (\partial G/\partial N_i)_{T, P, N_{k(k \neq i)}} \qquad (1.16)$$

$$= (\partial A/\partial N_i)_{T, V_m, N_{k(k \neq i)}}$$

Remembering that

$$(\partial n_i/\partial n)_{n_{k(k \neq i)}} = (\partial N_i/\partial N)_{N_{k(k \neq i)}} = 1$$

we obtain from (1.10) or (1.11)

$$\mu_i^{pg}(T, V)/RT = \ln(N/V) - 1 + \ln x_i$$

or

$$\left. \begin{array}{c} \\ \\ \end{array} \right\} \qquad (1.17)$$

$$\mu_i^{pg}(T, P)/RT = \ln(P/kT) + \ln x_i$$

The equations for μ_i^{pg} and G^{pg} contain $\ln(P/kT)$ and thus depend on the choice of units used in the calculations. All other terms, including those for real fluids, are dimensionless. However, when *phase equilibria* are calculated at constant T and P, the terms $\ln(P/kT)$ on both sides of the equilibrium condition (2.12) are identical and cancel each other. Hence, phase equilibria of pure substances and mixtures *can be evaluated without the knowledge of a standard or a reference pressure*.

Since the functions for a perfect gas are so simple and independent of the chemical nature of a gas, and since U^{pg}, H^{pg}, C_v^{pg}, and C_p^{pg} are independent of V, it is convenient to use them as reference functions and to operate with *differences between the properties of a real fluid and the perfect gas at the same temperature and pressure or volume*, called *the residual functions* and denoted by the superscript r. The configurational properties are now

$$Y(T, P, n_i) = Y^{pg}(T, P, n_i) + Y^r(T, P, n_i); \qquad (1.18)$$

$$Y(T, V, n_i) = Y^{pg}(T, V, n_i) + Y^r(T, V, n_i) \qquad (1.19)$$

Throughout this book we shall be concerned with the evaluation of configurational properties of real fluids.

The differences between $Y(T, P)$ and $Y(T, V)$ have been lucidly illustrated by Wilhelm (1974), and great care must be taken not to confuse the two kinds of functions. The residual functions were defined by Michels, Geldermans, and de Groot (1946) by the relations

$$Y^r(T, V) = \int_\infty^V \left[\left(\frac{\partial Y}{\partial V} \right)_T - \left(\frac{\partial Y}{\partial V} \right)_T^{pg} \right] dV$$

$$= \int_0^{1/V} \left[\left(\frac{\partial Y}{\partial(1/V)} \right)_T - \left(\frac{\partial Y}{\partial(1/V)} \right)_T^{pg} \right] d(1/V) \qquad (1.20)$$

$$Y^r(T, P) = \int_0^P \left[\left(\frac{\partial Y}{\partial P} \right)_T - \left(\frac{\partial Y}{\partial P} \right)_T^{pg} \right] dP \qquad (1.21)$$

It is often convenient to operate with densities $\rho = 1/V$ or molar densities $\rho_m = 1/V_m$ instead of volumes by using the transformation $\partial Y/\partial V = -\rho^2(\partial Y/\partial\rho)$.

For example, instead of

$$\frac{P}{NkT} = - \left[\frac{\partial(A/NkT)}{\partial V} \right]_T \qquad (1.22a)$$

we have

$$Z = \frac{P}{\rho_m RT} = \rho_m \left[\frac{\partial(A_m/RT)}{\partial \rho_m} \right]_T \qquad (1.22b)$$

Following Rowlinson (1958, 1969), the basic relations for the residual functions are derived from the *residual* compressibility factor, $(Z - 1)$. By substituting (1.22) into (1.20), we obtain the following equation:

$$\frac{A^r(T, V, n_i)}{nRT} = - \int_\infty^V (Z - 1) \frac{dV}{V} = \int_0^\rho (Z - 1) \frac{d\rho}{\rho} \qquad (1.23)$$

The residual internal energy is obtained from Eq. (1.23) by using the Gibbs-Helmholtz relation, which for the residual properties has the same form as Eq. (1.12):

$$\frac{U^r(T, V, n_i)}{nRT} = \int_\infty^V T \left(\frac{\partial Z}{\partial T} \right)_v \frac{dV}{V}$$

$$= - \int_0^\rho T \left(\frac{\partial Z}{\partial T} \right)_\rho \frac{d\rho}{\rho} \qquad (1.24)$$

The residual entropy is obtained from the second law, $TS^r(T, V) = U^r(T, V) - A^r(T, V)$. The residual heat capacity is obtained either from the definition $C_v^r = (\partial U^r/\partial T)_v$ or

$$\frac{C_v^r(T, V, n_i)}{nR} = T \int_\infty^V \left[2 \left(\frac{\partial Z}{\partial T} \right)_v + T \left(\frac{\partial^2 Z}{\partial T^2} \right)_v \right] \frac{dV}{V}$$

$$= - T \int_0^\rho \left[2 \left(\frac{\partial Z}{\partial T} \right)_\rho + T \left(\frac{\partial^2 Z}{\partial T^2} \right)_\rho \right] \frac{d\rho}{\rho} \qquad (1.25)$$

The residual functions $Y^r(T, P)$, defined by (1.21), are obtained analogously. For example, the familiar relation

$$G(T, P) = \int_0^P V \, dP,$$

at constant T, becomes

$$\frac{G^r(T, P, n_i)}{nRT} = \int_0^P (Z - 1) \frac{dP}{P} \tag{1.26}$$

Relations of this type, invariably appearing in all textbooks on thermodynamics, are seldom useful, because Z is usually expressed as a function of T and V and so may be differentiated or integrated with respect to V or ρ. The problem can be bypassed as follows.

Two of the configurational properties of the perfect gas derived above, A^{pg} and G^{pg}, depend on the volume of the system, and their variation must be considered simultaneously with that of the real fluid. Initially, the real gas is kept at T, V_1, and P_1 and the perfect gas at the same T and V_1 but at a different pressure because of the differences in Z of the two gases. We have certain values $A(T, V_1)$, $A^{pg}(T, V_1)$, and $A^r(T, V_1)$. In order to find the values of $A(T, P_1)$, and so on, where the real and the perfect gas are at the same pressure P_1, the volume of the latter has to be changed from V_1 to $V_2 = RT/P_1$. The corresponding free energy change is

$$\frac{A^{pg}(T, V_1) - A^{pg}(T, V_2)}{nRT} = - \ln \frac{V_1}{V_2} = - \ln \frac{V_1 P_1}{RT}$$
$$= - \ln Z \tag{1.27}$$

where Z is the compressibility factor of the real fluid. Another interesting relation follows from (1.10) and (1.11) when these functions are calculated at P_1 and V_1 of the real fluid:

$$\frac{G^{pg}(T, P, n_i)}{nRT} = \frac{A^{pg}(T, V, n_i)}{nRT} + \ln Z + 1 \tag{1.28}$$

By using the general relation $G(T, P) = A(T, V) + PV$, we obtain the desired equation in which $G^r(T, P)$ is expressed by the functions $Y^r(T, V)$:

$$\frac{G^r(T,\,P,\,n_i)}{nRT} = \frac{A^r(T,\,V,\,n_i)}{nRT} + Z - 1 - \ln Z \tag{1.29}$$

Since U^{pg} is independent of the volume, and the general thermodynamic equation for the enthalpy is $H(T, P) = U(T, V) + PV$, we obtain for the residual enthalpy

$$\frac{H^r(T,\,P,\,n_i)}{nRT} = \frac{U^r(T,\,V,\,n_i)}{nRT} + Z - 1 \tag{1.30}$$

Among the several known relations between the heat capacities $C_p(T, P)$ and $C_v(T, V)$, the most useful form is

$$C_p = C_v - T \left(\frac{\partial P}{\partial T} \right)_v^2 \left(\frac{\partial P}{\partial V} \right)_T^{-1} \tag{1.31}$$

which for the perfect gas simplifies to $C_p^{pg} = C_v^{pg} + R$. The heat capacities of the perfect gas are independent of the volume. Therefore, the relation between the residual functions is:

$$\frac{C_p^r(T,\,P,\,n_i)}{nR} = \frac{C_v^r(T,\,V,\,n_i)}{nR} - 1 - \frac{T}{R} \left(\frac{\partial P}{\partial T} \right)_v^2 \left(\frac{\partial P}{\partial V} \right)_T^{-1}$$

$$= \frac{C_v^r(T,\,V,\,n_i)}{nR} - 1$$

$$+ \frac{T}{R\rho^2} \left(\frac{\partial P}{\partial T} \right)_\rho^2 \left(\frac{\partial P}{\partial \rho} \right)_T^{-1} \tag{1.32}$$

The basic relation for the residual chemical potential of the *i*th component of a mixture is usually derived directly from (1.26) by using the definition (1.16), which in this case becomes

$$\mu_i^r = (\partial G^r / \partial N_i)_{T,\,P,\,N_k(k \neq i)} \tag{1.33}$$

and we obtain

$$\mu_i^r(T, P) = \int_0^P \left(\frac{P v_i}{RT} - 1 \right) \frac{dP}{P} \qquad \text{(constant } T\text{)} \tag{1.34}$$

where v_i is the partial molar volume of component i,

$$v_i = (\partial V/\partial N_i)_{T, P, N_k(k \neq i)}.$$

This relation has been widely used in chemical engineering, and great efforts were made to evaluate the partial molar volumes of the components of mixtures as functions of x_i by using graphical or analytical methods. If, however, an accurate equation of state of the mixture is known, another expression for μ_i^r becomes more useful. It is convenient to operate with quantities per mole of mixture, also called *mean molar quantities*, $Y_m = Y/n$, and the mole fractions $x_i = n_i/n$. For example, instead of (1.26), we have:

$$\frac{G_m^r(T, P)}{RT} = \int_0^P (Z_m - 1) \frac{dP}{P} \qquad \text{(constant } T) \qquad (1.35)$$

where $Z_m = PV_m/RT$ and V_m is the molar volume of the mixture.

As shown by Münster (1970) for an m-component system,

$$\mu_i(T, P) = G_m(T, P) - \sum_{\substack{j \neq i}}^{m-1} x_j \left[\frac{\partial G_m(T, P)}{\partial x_j} \right]_{P, T, x_k(k \neq i, j)} \qquad (1.36)$$

Analogously,

$$\mu_i(T, V) = A_m(T, V) - \sum_{\substack{j \neq i}}^{m-1} x_j \left[\frac{\partial A_m(T, V)}{\partial x_j} \right]_{T, V, x_k(k \neq i, j)} \qquad (1.37)$$

The difference between the two functions for the perfect gas is the same as in (1.28), so the fundamental relation between the residual functions is

$$\frac{\mu_i^r(T, P)}{RT} = \frac{\mu_i^r(T, V)}{RT} + Z_m - 1 - \ln Z_m$$

$$= \frac{A_m^r(T, V)}{RT} + Z_m - 1 - \ln Z_m$$

$$- \sum_{\substack{j \neq i}}^{m-1} x_j \left[\frac{\partial A_m^r(T, V)/RT}{\partial x_j} \right]_{T, V, x_k(k \neq i, j)} \qquad (1.38)$$

Remember that the derivative on the right-hand side is to be evaluated at *constant volume* and by keeping constant all the mole fractions other than x_i and x_j. Since μ_i^{pg} of the perfect gas depends on x_i, it must be taken into account when the equilibrium conditions are considered (that is, the equality of the chemical potentials of a given component in each of the phases in equilibrium at constant T and P; see Chap. 2). They require the equality of the configurational potentials given by

$$\frac{\mu_i(T,\,P)}{RT} = \ln\left(\frac{P}{kT}\right) + \ln x_i + \frac{\mu_i^r(T,\,P)}{RT} \tag{1.39}$$

where the last term is given by Eq. (1.38).

In practice, the relations will seldom be used for systems with more than three components. The following relations result from (1.39) for a ternary system:

$$\frac{\mu_1}{RT} = \ln\left(\frac{P}{kT}\right) + \ln x_1 + \frac{G_m^r}{RT} - x_2\left[\frac{\partial(A_m^r/RT)}{\partial x_2}\right]_{x_3} - x_3\left[\frac{\partial(A_m^r/RT)}{\partial x_3}\right]_{x_2};$$

$$\frac{\mu_2}{RT} = \ln\left(\frac{P}{kT}\right) + \ln x_2 + \frac{G_m^r}{RT} - x_1\left[\frac{\partial(A_m^r/RT)}{\partial x_1}\right]_{x_3} - x_3\left[\frac{\partial(A_m^r/RT)}{\partial x_3}\right]_{x_1};$$

$$\frac{\mu_3}{RT} = \ln\left(\frac{P}{kT}\right) + \ln x_3 + \frac{G_m^r}{RT} - x_1\left[\frac{\partial(A_m^r/RT)}{\partial x_1}\right]_{x_2} - x_2\left[\frac{\partial(A_m^r/RT)}{\partial x_2}\right]_{x_1};$$

We note that $(\partial/\partial x_2)_{x_3} = -(\partial/\partial x_1)_{x_3};\ (\partial/\partial x_1)_{x_2} = -(\partial/\partial x_3)_{x_2};\ (\partial/\partial x_2)_{x_1} = -(\partial/\partial x_3)_{x_1};$

so that the foregoing relations fulfill the thermodynamic condition for partial molar quantities:

$$\sum_{j \neq i} x_i x_j \left[\frac{\partial(A_m^r/RT)}{\partial x_j}\right]_{x_k(k \neq i,\,j)} = 0 \tag{1.40a}$$

In this case,

$$\frac{G_m}{RT} = \sum_i^m x_i\,\frac{\mu_i}{RT} = \ln\frac{P}{kT} + \sum_i^m (x_i \ln x_i) + \frac{G_m^r}{RT}$$

Eq. (1.40a) is a special case of the generalized Gibbs-Duhem equation (Münster, 1970):

$$\sum_{i}^{m} n_i \, dy_i - (\partial Y/\partial T)_{P, n} \, dT - (\partial Y/\partial P)_{T, n} \, dP = 0 \tag{1.40b}$$

where

$$y_i = (\partial Y/\partial n_i)_{T, P, n_j(j \neq i)}$$

and Y is an extensive function of state (configurational property).

The foregoing relations have been derived in recent years, and only some of them can be found in textbooks on thermodynamics. They are used in recent papers in which phase equilibria, particularly those at high pressures, are predicted from theoretical equations for the Helmholtz free energy and the compressibility factors.

The *fugacity* is a related function extensively used by chemical engineers in the United States. It was introduced by Lewis and Randall (1923) and has been exhaustively discussed and applied by Prausnitz (1969) and his coworkers. The fugacity f_i of component i is defined by

$$\mu_i(T, P_1) - \mu_i(T, P_2) = RT \ln \frac{f_i(T, P_1)}{f_i(T, P_2)} \tag{1.41}$$

For a pure fluid, μ_i is replaced by G. By subtracting the contribution of the perfect gas, given by Eq. (1.17), we obtain the residual functions (ln kT cancels in the subtraction process)

$$\ln \frac{f_i(T, P)}{Px_i} = \frac{\mu_i^r(T, P)}{RT} \tag{1.42}$$

The dimensionless ratio f_i/Px_i is called the *fugacity coefficient*. For the perfect gas or a perfect gas mixture, it equals unity because $\mu_i^r = 0$ (by definition). For a pure fluid, (1.42) reduces to

$$\ln \frac{f(T, P)}{P} = \frac{G_m^r(T, P)}{RT} \tag{1.43}$$

The ratio $f(T, P)/P$ is often called the *activity coefficient*. As shown in Chap. 2, the activity coefficient of a component of a mixture has another definition, although it is again simply related to μ_i^r and G^r. To avoid confusion, the activity coefficient defined by (1.43) should rather be called the *fugacity coefficient of a pure fluid*.

The identities (1.42) and (1.43) indicate to us that we may operate equally well with fugacity coefficients and residual potentials. Lewis introduced the concept of fugacity to make it easier to understand the meaning of the chemical potential. He also wanted a potential whose lower limit was zero instead of $-\infty$. Now, in view of (1.42), the situation appears to be the reverse. While μ_i^r is directly related to the intermolecular energy *and* has the dimension of energy, the fugacity has the dimension of pressure; it requires some abstract thinking to associate pressure with intermolecular energy.

By definition (1.16), the chemical potential is the differential change in the free energy of a system at constant T and P, or the change in G of a very large number of molecules when one molecule of one of the species present is added to the system while keeping the number of all other species constant. The useful work done by this molecule is different in the absence of intermolecular forces (μ_i^{pg}) from that done in their presence (μ_i), and the difference is equal to μ_i^r. The name *chemical potential* aptly reflects the fact that all the kinds of μ_i, except μ_i^{pg}, depend on the chemical nature of the given molecule *and* all the other molecules in the assembly.

The residual functions can be evaluated from experimental data when an accurate equation for A_m^r or Z_m is known. At the present time, we may consider the equation of state (1.2) of a gas at very low densities. As shown by Kirkwood (1935), for an m-component mixture of real gases,

$$Z_m = 1 + \rho_m \beta_m(T, x_i) = 1 + \rho_m \sum_{i=1}^{m} \sum_{j=1}^{m} x_i x_j \beta_{ij}(T)$$

$$= 1 + \frac{1}{V} \sum_{i=1}^{m} \sum_{j=1}^{m} n_i n_j \beta_{ij}(T) \qquad (\rho_m \to 0) \qquad (1.44)$$

From Eqs. (1.23) to (1.33), one obtains

$$\frac{A_m^r(T, V)}{RT} = \rho_m \beta_m \qquad (\rho_m \to 0); \qquad (1.45)$$

$$\frac{U_m^r(T, V)}{RT} = -\rho_m T \frac{\partial \beta_m}{\partial T} \qquad (\rho_m \to 0); \qquad (1.46)$$

$$\frac{C_v^r(T,\,V)}{R} = -\rho_m T\left[2\,\frac{\partial\beta_m}{\partial T} + T\,\frac{\partial^2\beta_m}{\partial T^2}\right] \qquad (\rho_m \to 0); \quad (1.47)$$

Since $\rho_m = P/RT$ in Eq. (1.44), one obtains

$$\frac{G_m^r(T,\,P)}{RT} = \frac{P}{RT}\,\beta_m \qquad (\rho_m \to 0); \tag{1.48}$$

$$\frac{H_m^r(T,\,P)}{RT} = \frac{P}{RT}\left[\beta_m - T\,\frac{\partial\beta_m}{\partial T}\right] \qquad (\rho_m \to 0); \tag{1.49}$$

$$\frac{C_p^r(T,\,P)}{R} = -\frac{PT}{R}\,\frac{\partial^2\beta_m}{\partial T^2} \qquad (\rho_m \to 0); \tag{1.50}$$

and from (1.36), for $\rho_m \to 0$,

$$\frac{\mu_i^r(T,\,P)}{RT} = \frac{P}{RT}\left[\beta_m - \sum_{j\neq i}^{m-1} x_j\left(\frac{\partial\beta_m}{\partial x_j}\right)_{T,\,P,\,x_{k(k\neq i,\,j)}}\right] \tag{1.51}$$

The foregoing functions become equal to zero in the limits $Y^r(T, V)$ when $\rho_m = 0$, and $Y^r(T, P)$ when $P = 0$. However, none of the derivatives $(\partial Y^r/\partial\rho_m)$ $(\rho = 0)$ or $(\partial Y^r/\partial P)$ $(P = 0)$ are equal to zero for an imperfect gas; thus it always differs from the perfect gas. Eq. (1.45) may be regarded as the thermodynamic definition of the second virial coefficient.[†]

The derivatives $(\partial/\partial x_i)_{x_{k(k\neq i,\,j)}}$ and $(\partial/\partial x_i)$ are as follows. If the function is that most commonly obtained in statistical mechanics, that is

$$Y_m = \sum_{i=1}^{m}\sum_{j=1}^{m} x_i x_j\,Y_{ij} \tag{1.52}$$

then, if all the mole fractions *except two* are kept constant,

$$(\partial Y_m/\partial x_i)_{x_{k(k\neq i,\,j)}} = Y_{ii} - Y_{jj} + (2Y_{ij} - Y_{ii} - Y_{jj})(x_j - x_i)$$

$$+ \sum_{k=1}^{m-2} x_k[2(Y_{ik} - Y_{jk}) + Y_{jj} - Y_{ii}] \tag{1.53}$$

[†]Tables of smoothed values of second virial coefficients, prepared by the author, were published by Zwolinski et al. (1970–74). The experimental values for pure gases were compiled by Dymond and Smith (1969) and for mixtures by Warowny and Stecki (1979).

If all the mole fractions are allowed to vary, then

$$\partial Y_m/\partial x_i = Y_{ii} - \sum_{j \neq i} Y_{jj} + \sum_{j \neq i} (2Y_{ij} - Y_{ii} - Y_{jj})(x_j - x_i)$$

$$- \sum_{k > j} (2Y_{jk} - Y_{jj} - Y_{kk})(x_j + x_k) \qquad (j, k \neq i) \quad (1.54)$$

In a binary system, $x_k = 0$, and (1.53) and (1.54) become identical because the terms with $(2Y_{jk} - Y_{jj} - Y_{kk})$ do not appear in (1.52).

Examples of the residual properties will be given in later chapters. At the present time it is proper to outline how these functions are evaluated from accurate experimental data for pure fluids. For this purpose the compiler applies an accurate empirical equation of state with many constants (often more than forty) and separate smoothing equations for the saturated vapor pressures and the densities of the coexisting liquid and vapor phases. The results are combined to obtain a PVT or $Z(T, V)$ or $Z(T, P)$ diagram of the substance. The Helmholtz free energies A^r are then obtained by an integration. However, the values of U^r are less accurate, as they are obtained by a differentiation. The values C_v^r can be only estimated in this way, as they correspond to the second derivatives. If, however, the directly measured values of C_p are known (even in a limited range of T and P), accurate values of H^r and S^r can be obtained by an integration procedure. The constants of the equation of state are then obtained by fitting them simultaneously to the observed values of Z and to H^r derived from the observed values of C_p. The procedures are best explained and illustrated by Din (1956).

Tables of accurate sets of the above thermodynamic properties have been constructed in recent years by Din (1956); by Zwolinski et al. (since 1968) as a part of American Petroleum Institute Project 44 and Texas A&M Thermodynamics Research Center research projects; by Angus, Armstrong, and de Reuck (since 1971) as a IUPAC project; by Gosman, McCarty, and Hust (1969); and by Goodwin et al. (since 1969) of the National Bureau of Standards. These tables differ in their manner of data presentation. For example, the values of U and H for argon are given by Angus and coworkers (1971) with respect to the ideal crystal at $T = 0$ with an additive constant $192.5197 \ J \cdot g^{-1}$. In this case,

$$\frac{U^r(T, V)}{RT} = \frac{(U^{\text{TAB}} - 192.5197) \ M}{RT} - \frac{3}{2} \qquad \text{(Argon)}$$

where U^{TAB} are the tabulated values in $J \cdot g^{-1}$ and M is the atomic mass of argon.

In the tables of thermodynamic properties of ethane, prepared by Goodwin and coworkers (1976), U^{TAB} represent total values in $J \cdot mol^{-1}$ with the zero-point energy taken as $U_o^o = 20211 \ J \cdot mol^{-1}$. They contain U^{mol} calculated by J. Chao and Zwolinski (1973).[†] In this case,

$$U^r/J \cdot mol^{-1} = U^{TAB} - U^{mol} - 20211 \qquad \text{(ethane)}$$

Additionally, a reference point for the entropy is chosen. Although the third law of thermodynamics suggests $S = 0$ for an ideal crystal at 0 Kelvin, other reference points have been used.

If $C_p(T, \rho)$ is known, the route usually followed is to establish $S(T, \rho)$ and $H(T, \rho)$ first. However, Din (1956) notes that it would be easier to evaluate first $H(T, \rho)$ and $G(T, \rho)$. The entropy $S(T, \rho)$ of a pure fluid would then be obtained from the second law of thermodynamics. The thermodynamic relations, particularly useful when an accurate equation of state of the fluid is known, are

$$A(T, V) = H(T, P) + T(\partial A/\partial T)_v - PV \qquad (1.55)$$

and the sum of Eqs. (1.28) and (1.29). In this case a reference value for H only must be chosen—say, that of the ideal crystal or the perfect gas at $T = 0$. These problems are more extensively treated in the excellent book by Bett, Rowlinson, and Saville (1975).

A common reference point for all substances will facilitate the use of the tables for pure fluids to calculate the "total" functions for mixtures of fluids. A convenient reference point is the state of the perfect gas at $T = 0$ and $P^o = 1$ atm $= 1.01325$ bar. The molecular functions H^{mol} and S^{mol}—denoted in the tables published by TRC by $(H - H^\circ)$ and S, respectively—in this state depend on the temperature only. If an accurate equation of state for the mixture is known, allowing the calculation of the residual functions at a given T and P, the "total" functions are

$$H^{total} = H^r + \sum_i x_i H_i^{mol} \qquad (1.56)$$

[†]The tables of thermodynamic molecular properties of perfect gases worked out by J. Chao and coworkers (since 1961) at the Thermodynamic Research Center are the most extensive and accurate existing sets of data.

and

$$S^{\text{total}} = S^r + \sum_i x_i S_i^{\text{mol}} - R \sum_i x_i \ln x_i - R \ln(P/P^{\oplus}) \qquad (1.57)$$

If the mixture is liquid, H^r calculated from an equation of state already includes the enthalpy of vaporization.

Less accurate but very useful in chemical engineering are the *generalized* tables and graphs of the thermodynamic functions. They are based on the extended principle of corresponding states with three parameters (P^c, T^c, and ω). The construction and applications of such tables are best explained by K. C. Chao and Greenkorn (1975) and by Starling (1973).

It is clear that the thermodynamic relations derived in this chapter for the residual functions are useful only when at least one of them, say Z, was measured or evaluated from a theory based on statistical mechanics and the theory of intermolecular forces. The other ones, difficult or impossible to measure directly, are then calculated from the thermodynamic relations. As stated by McGlashan (1965),

What do we learn from thermodynamics about the microscopic explanation of macroscopic changes? Nothing whatever. What then is a thermodynamic theory? There is no such thing. What then is the *use* of thermodynamics? Thermodynamics is useful precisely because some quantities are easier to measure than others and that is all.

CHAPTER 2

Thermodynamic Excess Functions

Since Raoult discovered his law, it has been convenient to compare the properties of real mixtures with those of an *ideal mixture* or *ideal solution*.[†]

The concept is similar but not identical to that upon which the residual functions are based. While in a perfect gas mixture the configurational energies $\mathcal{U}$ of all the components are zero, those in an ideal mixture are non-zero because such a system may be a mixture of *imperfect* gases or liquids.[‡]

Ideal mixtures are mixtures whose components have either identical molecular sizes and interaction energies or such similar residual Helmholtz energies that the molecular arrangements and the concentrations of components can be freely changed without affecting the free energy of the system. Isotopic mixtures are often given as examples of ideal mixtures, but as shown by Prigogine and coworkers (1957) and confirmed experimentally by Walters and Fairbank (1956), $^3He + {}^4He$ mixtures may separate into two liquid layers, which clearly points to a non-ideal behavior. This system, as well as those involving hydrogen or neon, is exceptional. There are a few other examples of very nearly ideal systems that have nonnegligible differences in $u(r)$ of the components, but it is doubtful that they remain ideal over a wide temperature range.

The molecular origin of ideal mixtures is clearly explained by Ben-Naim (1975) in terms of the pair distribution function $g(r)$. The quantity $\rho_n g(r) 4\pi r^2 dr$ is the average number of particles in a spherical shell of width dr at a distance r from the center of a chosen particle. If the center were chosen at random or the distribution of all the particles

---

[†] The thermodynamic relations for *mixtures* and *solutions* are identical. The term *solution* will be used only when one of the components is in large excess, and it is then called the *solvent*.

[‡] Perfect gases are often called *ideal* gases, and the term is not confusing as long as pure fluids are considered. However, *perfect mixtures* (mixtures of perfect gases) must be clearly distinguished from ideal mixtures.

were random, $g(r)$ would be equal to unity. The difference integrated over all distances is given by

$$F_{ij} = \int_0^\infty [g_{ij}(r) - 1]4\pi r^2 dr \tag{2.1}$$

For the perfect gas at all densities, $g(r) = 1$. For a real fluid at very low densities, $g(r)$ approaches the Boltzmann distribution law, $g(r) = \exp[-u(r)/kT]$, which leads to Eq. (1.3).

Kirkwood and Buff (1951) derived the general relation for $(\partial \mu_i / \partial \rho_{ni})_{T,P}$ of a component of a binary system in terms of F_{ij}. The relation can easily be converted to $(\partial \mu_i / \partial x_i)_{T,P}$ because the number densities ρ_{ni} and x_i are related: $x_1 = N_1/(N_1 + N_2) = \rho_{n1}/(\rho_{n1} + \rho_{n2}) = \rho_{n1}/\rho_n$. The result obtained by Ben-Naim is

$$\left(\frac{\partial \mu_1}{\partial x_1}\right)_{T,P} = RT\left[\frac{1}{x_1} - \frac{\rho_m x_2(F_{11} + F_{22} - 2F_{12})}{1 + \rho_m x_1 x_2(F_{11} + F_{22} - 2F_{12})}\right] \tag{2.2}$$

If the interactions between the components are identical, $F_{11} + F_{22} - 2F_{12} = 0$, and

$$(\partial \mu_i / \partial x_i)_{T,P} = RT/x_i \qquad (0 \leq x_i \leq 1)$$

By integration we obtain

$$\mu_i(T, P) = \text{const.} + RT \ln x_i \qquad (0 \leq x_i \leq 1) \tag{2.3}$$

The constant must be equal to the chemical potential of the pure component i, $G_i(T, P)$, because for $x_i = 1$, $\mu_i = G_i$. Hence, *for each component* of the ideal system,

$$\frac{\mu_i^{id}(T, P)}{RT} = \frac{G_i(T, P)}{RT} + \ln x_i \qquad (0 \leq x_i \leq 1) \tag{2.4}$$

This is a commonly used definition of an ideal mixture. By comparison with (1.17), it is clear that for a mixture of perfect gases the constant in (2.3) is equal to $\ln(P/kT)$. In contrast, the constant for an ideal mixture depends on the chemical nature of the component. Eq. (2.2) also expresses the fact that $\partial \mu_1/\partial x_1$ tends to RT/x_1 when $x_2 \to 0$ even if $(F_{11} + F_{22} - F_{12}) \neq 0$. Such a case is called the *ideal dilute solution* (of component 2 in this case). This name is misleading because μ_1 tends to be ideal, but $\partial\mu_2/\partial x_2$ of the component 2 in infinite dilution exhibits the greatest deviations from ideality when $x_1 \to 1$; $x_2 \to 0$.

For an ideal system it follows from (2.4), at constant T and P, that

$$\frac{G_m^{id}}{RT} = \sum_i x_i \frac{G_i}{RT} + \sum_i x_i \ln x_i; \tag{2.5}$$

$$\frac{H_m^{id}}{RT} = \sum_i x_i \frac{H_i}{RT}; \tag{2.6}$$

$$V_m^{id} = \sum_i x_i V_i \quad \text{and} \quad Z_m^{id} = \sum_i x_i Z_i; \tag{2.7}$$

$$C_{pm}^{id} = \sum_i x_i C_{pi} \tag{2.8}$$

The difference between (2.6) and (2.5) is S_m^{id}/R.

One of the most interesting relations for our purposes is the one between μ_i and the liquid-vapor equilibrium pressure and composition at constant pressure. Since the components of an ideal mixture have (nearly) identical molecular properties, their vapor pressures and those of the mixtures should be practically the same. However, the components of real systems usually have different vapor pressures. The deviations from an ideal mixture are calculated by using (2.4) as the reference equation with a constant G_i, whereas G_i in fact varies along the vapor pressure curve $P(T, x_i)$, according to the thermodynamic relation (Rowlinson, 1969)

$$\frac{\partial(G_i/RT)}{\partial \ln P} = \frac{PV_i}{RT} \tag{2.9}$$

where V_i is the liquid molar volume of pure component i. If the deviations from an ideal mixture are calculated *along the saturation curve* $P(T, x_i)$ (variable P), Eq. (2.4) is useful only at very low pressures of *all the components* when the compressibility factor of the liquid is very small, $P \to 0$.

The conditions of equilibrium between two or more phases were derived in 1875 by Gibbs (1948). With the second phase denoted by a prime, the third phase by a double prime, and so on, the conditions are

$$T = T' = T'';\tag{2.10}$$

$$P = P' = P'' \quad \text{or} \quad Z/V_m = Z'/V'_m = Z''/V''_m \tag{2.11}$$

and *for each component* of the system

$$\mu_i (x_i, x_j, \ldots, V_m) = \mu'_i(x'_i, x'_j, \ldots, V'_m)$$
$$= \mu''_i (x''_i, x''_j, \ldots, V''_m) \tag{2.12a}$$

Condition (2.10) is obvious.

Condition (2.11) is also called the hydrostatic condition. If the pressure is not uniform throughout the system—for example, under the influence of gravitational or centrifugal forces—condition (2.12a) must include the gravitational potential. As stated by Guggenheim (1967), in the presence of a gravitational field even the simplest possible kind of system must be considered as composed of a continuous sequence of phases, each differing infinitesimally from its neighbors.

Condition (2.12a) tells us that the mole fractions of the components and the volumes of each phase adjust themselves so that the work of putting in one molecule of *any* of the species present is the same in each phase. It seems physically obvious, because if it were easier for a molecule of species i, say, in the first phase to penetrate the phase (') than the phase ("), x_i in the phase (') would increase until the *internal pressure* exerted by the other molecules would stop this transfer.

For a pure fluid, Eq. (1.39) becomes $\mu_i = RT \ln(P/kT) + G_i^r$. Hence, for a pure fluid at constant T and P, the equilibrium conditions are (2.11), that is,

$$Z/V_m = Z'/V'_m = Z''/V''_m$$

and

$$A_m^r/RT + Z - \ln V_m = (A_m^r/RT)' + Z' - \ln V'_m = \ldots \tag{2.12b}$$

or in terms of fugacities defined by Eq. (1.42),

$$f = f' = f'' \tag{2.12c}$$

In the same paper, Gibbs derived the famous *phase rule*. According to the conditions (2.10), (2.11), and (2.12), if there were a single phase, the m-component nonreacting system would be completely described by $(m + 2)$ quantities $T, P, \mu_1, \mu_2, \ldots, \mu_m$. However, one of the chemical potentials is not independent. As the example of Eq. (1.39) for a ternary system shows, one of the potentials, say μ_1, is fixed for the given values of μ_2 and μ_3 by the condition that the sum of $x_i x_j (\partial A_m^r / \partial x_j)_{x_k} = 0$. Accordingly, a single phase has only $(m + 1)$ *degrees of freedom* of choice. If there are p phases in equilibrium, then the number of such restricting relations is p, and the number of degrees of freedom is $(m + 2 - p)$. It is presumed here that each of the components has different thermodynamic properties. As shown recently by R. L. Scott (1977), the rule should be modified for systems whose two or more components have *identical* thermodynamic properties: for example, mixtures of D- and L-optical enantiomers (called symmetric systems). In a symmetric binary system, as many as six phases may coexist at a given T and P. If such a system were treated as a one-component system, the phase rule would allow at most three phases in equilibrium. Except for this ambiguous case, the phase rule is helpful in the construction of phase diagrams. A system of one component and two phases has one degree of freedom. That is, either T or P may be changed, but there is only one curve $P_\sigma(T)$ for the liquid-vapor equilibrium. The lower end of this curve, the *triple point* at which three phases coexist (solid-liquid-vapor), and the upper end, the *critical point*, have zero degrees of freedom. They are fixed points for a given substance. At the triple point, each of the phases has a different molar volume. Thus, each of the terms A_m^r / RT, Z, and $\ln V_m$ in (2.12b) has a different value for each phase, but the sums $G_i = \mu_i$ are identical as required by (2.12a). At the critical point, the three phases (liquid-vapor-supercritical gas) have identical molar volumes; hence there is only one phase at this point. It is a fixed point because of two additional constraints on the system: namely, the critical point conditions $(\partial P / \partial V)_T = (\partial^2 P / \partial V^2)_T = 0$ (Chap. 7). Thus, the number of degrees of freedom is (*see*, for example, Morrison, 1981):

$$\text{d.f.} = m + 2 - (p + c) \qquad (2.13)$$

where c is the number of additional constraints.

A binary system has one degree of freedom more than a pure fluid because of $\mu_i(x_1)$. A single $P_\sigma(T)$ curve is obtained here for three phases in equilibrium (liquid-liquid-vapor), each with different values of V_m and x_1. The fixed points with d.f. $= 0$, at which this curve ends, are the *quadruple point* (solid-liquid-liquid-vapor) and the *critical end points* with two phases in equilibrium and $c = 2$ (Chap. 7).

The quantities usually measured in the studies of phase equilibria are the pressures of the system and the mole fractions of the components x_i, x_i', and x_i'' in the fluid phases at a constant temperature. Under certain conditions, x_i'' may become identical with x_i or x_i'. We shall denote by x_i' the mole fractions in the vapor phase. (In the case of a gas-gas equilibrium, x_i' is for the phase rich in the more volatile component. For a one-phase system, x_i without the prime will be used—for example, Eq. [1.44].) The vapor pressure of the liquid phase in equilibrium with an infinitely small amount of the vapor phase is called the *total pressure* or the *bubble point* at the given x_i and T. The pressure at which the liquid phase very nearly vanishes is called the *dew point* at the given x_i and T. Analogous curves as functions of x_i, obtained under isobaric (constant P) conditions, are sometimes called the *boiling temperatures* and the *condensation temperatures*, respectively, at constant pressure.

Gibbs's equilibrium conditions, when applied to an ideal mixture at *very* low pressures, lead to Raoult's law. At low pressures, $(\mu_i^r)'$ in the vapor phase becomes so small that we can apply Eq. (1.17) for the perfect gas; that is, $\mu_i' = (\mu_i^{pg})'$. Initially, all the *pure* liquids are in equilibrium with their vapors at the saturated vapor pressures P_i ($x_i = x_i' = 1$; $\ln x_i = 0$). Consider the changes $\Delta\mu_i$ and $\Delta\mu_i'$ when all the liquids are mixed at constant T. The new equilibrium pressure is P, and each $x_i < 1$, $x_i' < 1$. For each component in the liquid phase,

$$\Delta\mu_i = \mu_i(T, P_i) - \mu_i(T, P) = \Delta G_i - RT \ln x_i$$

and in the vapor phase,

$$\Delta\mu_i' = \mu_i'(T, P_i) - \mu_i'(T, P) = RT \ln(P_i/P) - RT \ln x_i'$$

where, according to (2.9), ΔG_i is negligibly small and equal to $(P_i - P)V_i$, where V_i is the liquid molar volume of the component i. Further, $\Delta\mu_i = \Delta\mu_i'$ because both the initial state and the final state are equilibrium states. Hence the dew-point composition is

$$x_1' = (x_1 P_1/P) \exp(-\Delta G_1/RT);$$

$$x_2' = (x_2 P_2/P) \exp(-\Delta G_2/RT); \text{ etc.}$$

But $x_1' + x_2' + \ldots + x_m' = 1$; therefore, the total or bubble-point pressure is

$$P = \sum_i^m x_i P_i \exp[(P - P_i)V_i/RT] \approx \sum_i^m x_i P_i \tag{2.14}$$

which is Raoult's law.

Nearly all real mixtures are non-ideal. The concept of an ideal system is useful only as a reference system for the *thermodynamic excess functions*, first proposed by Scatchard (1937). An excess function is the difference between the given function of a real mixture and that of an ideal mixture at the same T, P, and x_i (but not ρ or ρ^*). The excess functions will be marked by the superscript E. Accordingly, if μ_i is the configurational chemical potential of a component of a real mixture, then at constant T, P, and x_i,

$$\frac{\mu_i^E(T, P)}{RT} = \frac{\mu_i(T, P) - G_i(T, P)}{RT} - \ln x_i \qquad (2.15)$$

By substituting (1.39) into (2.15), we obtain the basic relation for the excess chemical potential of the ith component of an m-component system:

$$\frac{\mu_i^E(T, P)}{RT} = \frac{A_m^r(T, V)}{RT} + Z_m - \ln V_m$$

$$- \sum_{j \neq i} x_j \left[\frac{\partial (A_m^r/RT)}{\partial x_j} \right]_{x_{k(k \neq i, j)}}$$

$$- \frac{A_i^r(T, V)}{RT} - Z_i + \ln V_i \qquad (2.16)$$

and for the excess Gibbs energy:

$$\frac{G^E(T, P)}{RT} = \sum_i x_i \frac{\mu_i^E}{RT}$$

$$= \frac{A_m^r(T, V) + PV^E}{RT} - \ln V_m$$

$$- \sum_i x_i \left(\frac{A_i^r(T, V)}{RT} - \ln V_i \right) \qquad (2.17)$$

where

$$V^E = V_m - \sum_i x_i V_i \qquad (2.18)$$

is the excess volume. The thermodynamic relations between the excess functions are the same as between the total functions. For example, the Gibbs-Helmholtz relation for the excess enthalpy is

$$\frac{H^E(T, P)}{RT} = -T\left[\frac{\partial(G^E/RT)}{\partial T}\right]_p \tag{2.19}$$

and for the excess internal energy,

$$\frac{U^E(T, V)}{RT} = -T\left[\frac{\partial(A^E/RT)}{\partial T}\right]_v \tag{2.20}$$

The function H^E can also be expressed as the following function of U^E:

$$\begin{aligned}
\frac{H^E(T, P)}{RT} = &\frac{U_m^r(T, V)}{RT} - \sum_i x_i \frac{U_i^r(T, V)}{RT} \\
&+ \frac{P}{R}\left[\frac{V^E}{T} - \left(\frac{\partial V^E}{\partial T}\right)_{P, x} + \frac{1}{Z_m}\left(\frac{\partial V_m}{\partial T}\right)_{P, x}\right. \\
&\left. - \sum_i \frac{x_i}{Z_i}\left(\frac{\partial V_i}{\partial T}\right)_{P, x}\right]
\end{aligned} \tag{2.21}$$

Obviously, the excess entropy

$$S^E/R = (H^E - G^E)/RT \tag{2.22}$$

and the excess heat capacity

$$C_p^E = (\partial H^E/\partial T)_p \tag{2.23}$$

If the excess functions are calculated from experimental data *along the saturation curve*, the results are accurate only at $PV_i/RT \rightarrow 0$ for the liquid phase (Eq. 2.9). At $P = 0$, the foregoing relations become

$$\begin{aligned}
\frac{\mu_i^E(T, P)}{RT} = &\frac{A_m^r(T, V)}{RT} - \ln V_m - \sum_{j \neq i} x_j \left[\frac{\partial(A_m^r/RT)}{\partial x_j}\right]_{x_{k(k \neq i, j)}} \\
&- \frac{A_i^r(T, V)}{RT} + \ln V_i \quad (P = 0); \tag{2.24}
\end{aligned}$$

$$\frac{G^E(T, P)}{RT} = \frac{A^E(T, V)}{RT} = \frac{A_m^r(T, V)}{RT} - \ln V_m$$

$$- \sum_i x_i \left(\frac{A_i^r(T, V)}{RT} - \ln V_i \right) \quad (P = 0); \quad (2.25)$$

$$\frac{H^E(T, P)}{RT} = \frac{U^E(T, V)}{RT}$$

$$= \frac{U_m^r(T, V)}{RT} - \sum_i x_i \frac{U_i^r(T, V)}{RT} \quad (P = 0) \quad (2.26)$$

Relations (2.24) to (2.26) are particularly useful when the excess functions are calculated from an equation of state, $P(T, V, x_i)$, or a theoretical equation for $A_m^r(T, V, x_i)$. Relations (2.17) and (2.21) are useful for the experimentalist who was forced to measure G^E and H^E at pressures greatly exceeding the condition of zero pressure and desires to convert the results to those that would be obtained at $P = 0$. The differences between (2.25) and (2.17) are negligible at pressures not exceeding about 5 bar.

Relations (2.15) to (2.20), *applied at constant P*, are not limited to low pressures. They can be applied to compressed liquid mixtures at a constant P greater than any saturation pressure in the range $0 \leq x_i \leq 1$. An example is shown later, in Table 7.2. Eq. (2.16) can also be written as

$$\mu_i^E(T, P) = \mu_i^r(T, P) - G_i^r(T, P) \quad (2.27)$$

which provides the physical meaning of μ_i^E. It is the difference between the work done when putting one molecule of species i into the mixture and that done upon adding the same molecule to the pure fluid i. That is, μ_i^E depends on the differences between the intermolecular forces of species i and those of the other species present in the mixture. If they are the same, then $\mu_i^E = 0$ for each of the components (an ideal mixture). At infinite dilution of species i, μ_i^E is the largest (or the smallest if the deviations are negative) because the single molecule i encounters the largest concentration of all other molecules with different intermolecular forces. The variation of μ_i^E and of the other excess functions with x_1 in a binary system are shown schematically in Fig. 2.1. The sign of the excess functions depends on molecular inter-

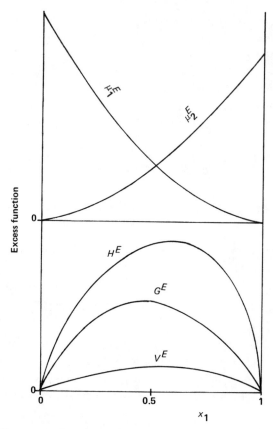

FIG. 2.1.   The excess functions of a binary system (the case of positive values) as functions of x_1 at constant T and P. The scale (and the units for V^E) are arbitrary.

actions that will be discussed later. Fig. 2.1 represents a common case when all the functions are positive.

A popular measure of the deviations from an ideal mixture is the *activity coefficient* γ_i of component i, which is defined as

$$\ln \gamma_i = \mu_i^E(T, P)/RT \tag{2.28}$$

The definition is analogous to that of the fugacity coefficient, (1.42), but the physical meanings of the two are different.

The relation between the total vapor pressure P of a non-ideal system and μ_i^E and P_i of its components, which we shall now derive, forms the most important bridge between experimental data and a theory of mixtures. We proceed along the same route as in the derivation of Raoult's law, (2.14). When other components are added to the pure fluid i, the pressure (as before, very low) changes from P_i to P, and the mole fraction x_i from $x_i = 1$ to a certain value $x_i < 1$. The chemical potential $\mu_i = \mu_i^{id} + \mu_i^E$ in the liquid phase changes at constant T by

$$\frac{\Delta \mu_i}{RT} = \frac{\mu_i(T, P_i) - \mu_i(T, P)}{RT}$$

$$= \frac{(P_i - P)V_i}{RT} - \ln x_i - \frac{\mu_i^E}{RT} \tag{2.29}$$

For a condensed phase, the variation of μ_i^E with the pressure is very weak and was neglected in the above relation.

The vapor phase is now an imperfect gas. If it were an ideal mixture, all the second virial coefficients would be identical, $\beta_{11} = \beta_{22} = \dots$, and $\partial \beta_m / \partial x_i = 0$ (identical intermolecular energies). In this case, $\Delta \mu_i'$ in the vapor phase would be

$$\frac{\Delta \mu_i'}{RT} = \frac{(P_i - P)\beta_{ii}}{RT} + \ln \frac{P_i}{P} - \ln x_i' \qquad \text{(ideal system, } \rho_m \to 0)$$

However, the vapor is a non-ideal mixture to the same extent as is the liquid phase. From the definition (2.27), expressed by (1.48) and (1.51), we obtain for $\rho_m \to 0$

$$\frac{(\mu_i^E)'(T, P, x_i')}{RT} = \frac{P}{RT} \left[\beta_m - \beta_{ii} \right.$$

$$\left. - \sum_{j \neq i}^{m-1} x_j' \left(\frac{\partial \beta_m}{\partial x_j} \right)_{T, \, x_k'(k \neq i, \, j)} \right] \tag{2.30}$$

and

$$\frac{\Delta \mu_i'}{RT} = \frac{(P_i - P)\beta_{ii}}{RT} + \ln \frac{P_i}{P} - \ln x_i' - \frac{(\mu_i^E)'}{RT} \tag{2.31}$$

Since the two phases are in equilibrium, by equating (2.29) and (2.31), we obtain the composition of the vapor phase:

$$x_i' = \frac{x_i P_i}{P} \exp\left[\frac{\mu_i^E - (\mu_i^E)'}{RT}\right] \exp\left[\frac{(P - P_i)(V_i - \beta_{ii})}{RT}\right] \quad (2.32)$$

But $x_1' + x_2' + \ldots + x_m' = 1$. Therefore, the total pressure is

$$P = \sum_i x_i P_i \exp\left(\frac{\mu_i^E}{RT}\right) \exp\left[\frac{(P - P_i)(V_i - \beta_{ii})}{RT}\right] \exp\left[\frac{-(\mu_i^E)'}{RT}\right]$$

$$= \sum_i x_i P_i \gamma_i \exp\left[\frac{(P - P_i)(V_i - \beta_{ii})}{RT}\right] \exp\left[\frac{-(\mu_i^E)'}{RT}\right] \quad (2.33)$$

Eqs. (2.32) and (2.33) are the basic relations between the measured quantities P, P_i, x_i, and x_i' and the chemical potentials of the components that eventually may be calculated from a theory of mixtures. In the simple case of a binary system,

$$\beta_m = (x_1')^2 \beta_{11} + (x_2')^2 \beta_{22} + 2\,x_1' x_2' \beta_{12}$$

$$= x_1' \beta_{11} + x_2' \beta_{22} + x_1' x_2' b_{12} \quad (2.34)$$

where $b_{12} = 2\beta_{12} - \beta_{11} - \beta_{22}$. Therefore,

$$\frac{(\mu_1^E)'}{RT} = \frac{P}{RT}\,(x_2')^2 b_{12};$$

$$\frac{(\mu_2^E)'}{RT} = \frac{P}{RT}\,(x_1')^2 b_{12} \qquad (\rho_m \to 0) \quad (2.35)$$

Hence, for a binary system,

$$P = x_1 P_1 \gamma_1 \exp\left[\frac{(P - P_1)(V_1 - \beta_{11}) - P(x_2')^2 b_{12}}{RT}\right]$$

$$+ x_2 P_2 \gamma_2 \exp\left[\frac{(P - P_2)(V_2 - \beta_{22}) - P(x_1')^2 b_{12}}{RT}\right] \quad (2.36)$$

This relation was derived by Barker (1953). If experimental data for the bubble-point curve $P(x_i)$ at constant T are given and these pressures are low (less than about 5 bar), $\mu_i^E(P, x_i)$ of the components and the dew-point curve $P(x_i')$ of this system are calculated as follows. Initially, we put $(\mu_i^E)' = 0$ in Eq. (2.33) because x_i' is unknown. The initial value of $\gamma_i(P, x_i)$, thus obtained, is used to obtain from (2.32) an initial value of the dew point x_i' at the same pressure P (Fig. 2.2). This allows us to evaluate $(\mu_i^E)'$ and to obtain a better value of γ_i from (2.33). The iterations are repeated for each experimental point. The resulting values of $\mu_i^E(P, x_i)$ for the liquid phase are expressed by one of the empirical smoothing equations, discussed in Chap. 8, with constants b_1, b_2, etc. The program combines the above iterations with a least-squares method for evaluation of these constants. A large set of P, x_i, x_i', and μ_i^E values is thus reduced to a few constants b_i.

In absence of strong interactions between one or more of the components in the vapor, $(\mu_i^E)'/RT$ in (2.33) can be neglected. At temperatures at which the vapor pressure of a liquid is very low, β_{ii} is always negative, and the liquid molar volume V_i is small compared to $-\beta_{ii}$. Hence the sign of the term in brackets depends on that of $(P - P_i)$.

The total effect of all these corrections on P is small. By far the strongest effect is that of γ_i in the liquid phase. The effect of positive μ_i^E/RT (such as in Fig. 2.1) is shown in Fig. 2.2 for two cases of a binary system at a constant T: one when the components have similar vapor pressures, and one when they are very different. When P_i of the components are similar and μ_i^E/RT is large enough, an *azeotrope* may be formed (a *positive azeotrope* in this case). They are compared with the curves of ideal systems with the same values of P_i. The bubble-point curves are always above the dew-point curves. For a given x_1 at the same P and T, the mole fraction in the vapor x_1' is shifted toward the more volatile component—in this case, $P_1 > P_2$ and $x_1' > x_1$—or toward the positive azeotropic point; that is,

$$x_1' > x_1 \quad \text{when} \quad 0 < x_1 < \bar{x}_1;$$

$$x_1' < x_1 \quad \text{when} \quad \bar{x}_1 < x_1 < 1$$

where $\bar{x}_1$ is the azeotropic composition. In the case of a *negative azeotrope* (a minimum on the $P_\sigma(x)$ curve, $\mu_i^E < 0$), the inequalities are reversed. At the azeotropic point, $(\partial P/\partial x_i)_{T,\sigma} = 0$, and it follows from thermodynamics that the compositions of the phases must be identical, $\bar{x}_1 = \bar{x}_1'$, because the densities are not (Gibbs-Konowalow laws; *see* Rowlinson, 1969). The value of $\bar{x}_1$ may vary with temperature, and the *azeotropic line* $\bar{P}(T, \bar{x}_i)$ of a binary system with two

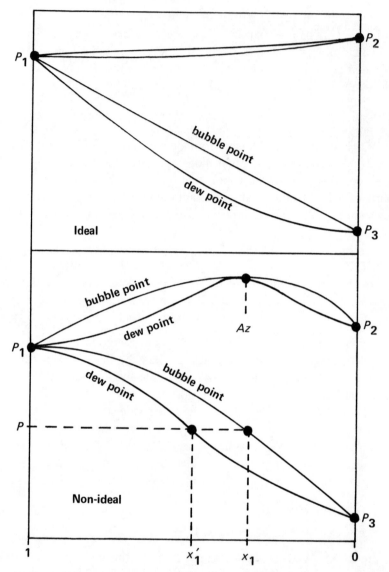

FIG. 2.2.   A comparison of the liquid-vapor equilibrium curves of two ideal and two non-ideal systems ($\mu_i^E > 0$) at a constant temperature. P_1, P_2, and P_3 are the saturation pressures of the three pure components.

phases has one degree of freedom: $c = 1$ in Eq. (2.13) because of the condition $(\partial P/\partial x_i)_{T,\sigma} = 0$. It may be a line of maximum points (see Fig. 4.1) or minimum points (Fig. 9.2) or both (as in the rare case shown in Fig. 9.1).

In the experimental studies of the equilibria at high pressures, both the bubble- and the dew-point curves are usually determined. At low pressures, however, the dew-point line is usually calculated from μ_i^E of the components obtained from the measured bubble-point curve. The curve so obtained is often more accurate than the directly measured compositions of the vapor. The reasons clearly follow from the shape of a *P-V* isotherm at a constant composition of the mixture, as shown in Fig. 2.3. The dew point is marked by the discontinuity of the *P-V* isotherm much less sharply than is the bubble point. For this reason, in most of the modern measurements at low pressures, only the total *P* as a function of x_i in the liquid phase is measured. Separate calorimetric measurements of H^E are made to determine $(\partial \mu_i^E/\partial T)_x$. At high pressures, the dew points, if not measured, may be calculated from known bubble points only by means of an accurate equation of state. Examples of such calculations are given in Chap. 7.

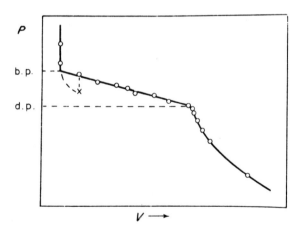

FIG. 2.3. Isothermal pressure-volume diagram of a mixture of constant composition, illustrating the determination of the bubble point (b.p.) and the dew point (d.p.). The cross represents a momentary superheating at the beginning of the expansion. Reproduced with permission from *The Characterization of Chemical Purity*, ed. L. A. K. Staveley, (London: Butterworth, 1971), p. 60.

Nevertheless, the knowledge of the excess functions, particularly G^E and H^E, is very important in the calculations of phase equilibria at *high* pressures. The reason is simple. Suppose that we have an accurate equation of state, valid at high pressures, in which the interaction parameters have a clear physical meaning: that is, an equation with a good theoretical foundation. It contains an expression for $u(r)$ with $u_{ij}(r)$—for mixed interactions—that we can predict in a few cases only. Now, the relations for A^r_m and U^r_m, derived from this equation of state, are substituted into (2.25) and (2.26). The values of $u_{ii}(r)$ are known from the properties of the pure components. The value of $u_{ij}(r)$ is varied until the calculated G^E and H^E curves agree with the experimental data. The equations are simple and permit rapid calculations. Then the same value of $u_{ij}(r)$ of the given system can be used to calculate the phase equilibria at high pressures by means of the equation of state. It is a long way from low pressures to the gas-liquid critical locus curve, and the dependence of $u(r)$ on the temperature must be properly estimated to obtain good results. We shall repeatedly return to this problem in the next chapters.

CHAPTER 3

Intermolecular Forces

The departures of the properties of all substances from those of the perfect gas are clear evidence of intermolecular attraction. The energy of interaction between molecules with static multipole moments (dipole, quadrupole, and so on), averaged over all orientations, is negative (attraction). However, the energy between small spherical molecules such as argon, calculated by means of classical physics, is positive because of the repulsion between the electronic shells. The attraction between argonlike molecules was a paradox until the development of quantum mechanics. Wang (1927) was the first who showed that the instantaneous dipole interactions between the stationary protons and the circling electrons have a negative sign and thus correspond to an attraction. Eisenschitz and London (1930) developed the theory further, and London (1930) obtained the following famous formula for the "dispersion" energy, which we shall call the *London energy*. The energy $u(r)$ between an *isolated* pair of small molecules, indicated by the subscript *pij* (*p* for an isolated pair), is

$$u^L_{pij}(r) = -\frac{3}{2}\frac{p_i p_j}{r^6}\left[\frac{1}{h\nu_i} + \frac{1}{h\nu_j}\right]^{-1} \tag{3.1}$$

where p_i and p_j are the mean polarizabilities (the usual symbol α_i is replaced by p_i to avoid confusion with the symbols used in the equations of state); $h\nu$ is the characteristic energies of the molecules; and r is the distance between the molecular centers. London supposed that $h\nu$ is equal to the ionization energy I, and Pitzer (1959) showed that it is in fact so for two-electron systems (H_2, He). For many-electron systems, the ratio $h\nu/I$ is greater than two and increases with the size of the molecules. Therefore, in deriving any combining rules for mixtures based on (3.1), it is best to eliminate the unknown quantities $h\nu$.

The formula for London energy, resulting from the theory of Slater and Kirkwood (1931), has similar limitations. More recently,

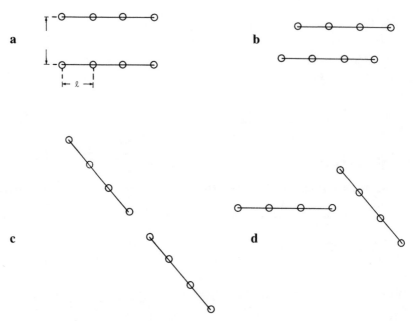

Fɪɢ. 3.1.   Schematic representation of four mutual orientations of two chain molecules or parts of chains of such molecules.

Linder (1960–64) obtained more general relations which, when simplified for practical applications, become analogous to (3.1).

The problem of London energies between large molecules has been clarified to some extent by the studies of interactions between chain molecules. First, the interaction energy depends on the mutual orientation of the chains. As shown by Davies and Coulson (1952), the London forces are strongest when the two chain molecules are parallel and opposite (Fig. 3.1a), and weaker when they are parallel and displaced (Fig. 3.1b); they vanish almost completely when the molecules are rotated in plane to the positions shown in Figs. 3.1c and 3.1d. From this study it appears that there are strong forces tending to align the molecules in positions parallel and opposite to one another. These calculations were performed for intermolecular distances greater than the length of the molecules.

Salem (1962) studied the interactions between n-alkanes at intermolecular distances that were small compared to the total length of the chain L but large relative to the length of a unit l (Fig. 3.1a). The centers of two units interact according to London Eq. (3.1), which for identical units can be written $u(r) = u^o/r^6$, where $u^o = -3/4 \, p^2 h\nu$. The total London energy between two parallel chains W_p (p for pair), each of length L and containing m units ($L = ml$), at a distance D between the center lines, is

$$W_p = \frac{u^o m}{4lD^5} \left[3 \tan^{-1}(L/D) + \frac{L/D}{1 + (L/D)^2} \right] \tag{3.2}$$

At large distances D, small L/D, $\tan^{-1}(L/D) \approx L/D$, and

$$W_p \approx (u^o/D^6)m^2 \qquad (L/D \to 0) \tag{3.3}$$

At distances D much smaller than L, $\tan^{-1}(L/D) \approx \pi/2$, and

$$W_p \approx u^o \, \frac{3\pi m}{8lD^5} \qquad (D \ll L) \tag{3.4}$$

For the interactions between two identical *circular chains* assumed to lie in parallel planes (sandwich configuration) at a distance D, Salem obtained a relation that is different from (3.2), but it simplifies for $D \ll L$ to the Eq. (3.4); $L = ml$ is the circumference of the circles in this case. Eq. (3.4) may be particularly useful for alkanes and cycloalkanes at high densities of the fluid where the average distance $D < L$.

London and other interactions (except the most recent work) are extensively considered in the excellent books by Hirschfelder, Curtiss, and Bird (1954) and by Margenau and Kestner (1969). The problem of triplet (three-body) contributions to the London forces is also reviewed by Fitts (1966) and Dalgarno (1967). For reasons explained later, we shall neglect triplet contributions to the intermolecular energy.

Besides London interactions, many molecular species exhibit permanent (static) *multipole moments* that are a source of additional attraction. Among the electrostatic contributions the most significant are the dipole and the quadrupole interactions. For "ideal" dipoles—that is, dipoles in which the distance between the charges is negligible compared to r—and for "ideal" quadrupoles with cylindrical symmetry,

$$u_{pij}^{\mu\mu}(r) = -\frac{\mu_i \mu_j}{r^3} [2 \cos \theta_i \cos \theta_j$$
$$- \sin \theta_i \sin \theta_j \cos(\phi_i - \phi_j)]; \tag{3.5}$$

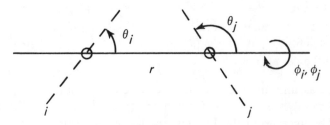

FIG. 3.2.   The four angles describing the mutual orientation of two molecules are shown by the dashed lines; ϕ_i and ϕ_j describe the rotation with respect to the *r*-axis.

$$u_{pij}^{\mu Q}(r) = \frac{3\mu_i Q_j}{4\ r^4} [\cos\ \theta_i\ (3\cos^2\ \theta_j - 1)$$
$$- 2\sin\ \theta_i\ \sin\ \theta_j\ \cos\ \theta_j\ \cos(\phi_i - \phi_j)]; \qquad (3.6)$$

$$u_{pij}^{QQ}(r) = \frac{3Q_i Q_j}{16\ r^5} \{1 - 5\cos^2\ \theta_i - 5\cos^2\ \theta_j - 15\cos^2\ \theta_i \cos^2\ \theta_j$$
$$+ 2[\sin\ \theta_i\ \sin\ \theta_j\ \cos(\phi_i - \phi_j)$$
$$- 4\cos\ \theta_i\ \cos\ \theta_j]^2\} \qquad (3.7)$$

where the angles θ and ϕ describe the mutual orientation of the dipoles or quadrupoles as shown in Fig. 3.2.

The interaction energies μ-Q and Q-Q between two isolated molecules may have either sign. However, the so-called effective spherically symmetrical potential functions, or the average interaction energy between a pair of molecules in a *large assembly of molecules*, are different from the ones above. They will be further indicated by the subscript *ij* (without *p*). The statistical averages[†] are

$$u_{ij}^{\mu\mu}(r) = -\mu_i^2\mu_j^2/(3kTr^6); \qquad (3.8)$$

$$u_{ij}^{\mu Q}(r) = -\mu_i^2 Q_j^2/(kTr^8); \qquad (3.9)$$

$$u_{ij}^{QQ}(r) = -7Q_i^2 Q_j^2/(5kTr^{10}) \qquad (3.10)$$

All the foregoing average interactions are then negative. Higher multipoles (octopoles, hexadecapoles) are discussed by Kielich (1965), Stogryn and Stogryn (1966), and Kihara (1970). The best reviews of

---

[†]The concept of the statistical average is due to Keesom (1921). Eqs. (3.8) to (3.10) are those given by Stell, Rasaiah, and Narang (1974).

experimental methods of determination of μ and Q are due to Buckingham (1959, 1970). A number of useful models of quadrupolar and octopolar molecules are demonstrated in the recent book by Kihara (1976).

At very low temperatures, the majority of the dipoles or quadrupoles assume the most favored relative orientations, shown in Fig. 3.3. At higher temperatures the orientations are increasingly chaotic.

The average of the London energy u_{ij}^L between two chain molecules, which would follow from (3.2), is yet unknown. It is certainly a function of the temperature, as it depends on the mutual orientation of the molecules. The dependence on r may change from r^{-5} for an isolated pair to r^{-10} (squared analogously to the multipole interactions) for a pair in a large assembly.

Fluids whose intermolecular forces depend on the mutual orientations of the molecules are grouped together as the *fluids with noncentral forces*. The most important contributions to the theory of noncentral forces were made by Keesom (1921), Cook and Rowlinson (1953), Pople (1954), Zwanzig (1955), Rowlinson and Sutton (1955), Stell, Rasaiah, and Narang (1972, 1974), and Gubbins and his coworkers. The work of Rowlinson and Gubbins and their coworkers, which yields practical solutions, is considered in more detail in the next chapter.

The theory of noncentral forces was recently confirmed for the case of London forces. Using the molecular beam-scattering techniques, Schmidt and Guillory (1976) determined the values of

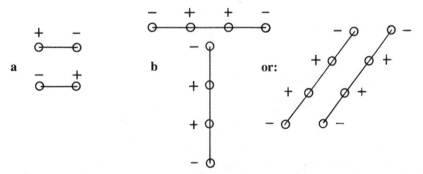

FIG. 3.3.   Favored mutual orientations of (a) two dipoles and (b) two quadrupoles with cylindrical symmetry. After Kirouac and Bose (1973).

TABLE 3.1.   The ratios of the force constants $\bar{u}_{ij}\sigma_{ij}^6$ of mixtures to that for methane, $\bar{u}_{ii}\sigma_{ii}^6$, determined by Schmidt and Guillory (1976).

System	$t/°C$	Ratio	System	$t/°C$	Ratio
$CH_4 + C_2H_6$	25	1.66	$CH_4 + n\text{-}C_4H_{10}$	25	3.67
	−60	1.72		−20	3.58
	−100	1.74		−40	3.61
	−130	1.82		−60	4.31
	−150	2.03		−80	4.83
$CH_4 + C_3H_8$	25	2.49	$CH_4 + n\text{-}C_5H_{12}$	29	5.09
	−43	2.66		0	5.30
	−73	2.93		−25	5.55
	−100	3.01		−50	6.18
	−130	3.10		−75	6.83

The uncertainties of these values range from ± 0.08 to ± 0.50. The average uncertainty is ± 0.2.

$\bar{u}_{ij}\sigma_{ij}^6/\bar{u}_{ii}\sigma_{ii}^6$ (where $\bar{u}$ are the mimimum values of $u(r)$ and σ are the collision diameters) of mixtures of *n*-alkanes with methane (*ij*) relative to that for methane (*ii*). The values are given in Table 3.1. It is clear that if $\bar{u}_{ij}$ varies with the temperature, then $\bar{u}_{jj}$ of the pure *n*-alkane must vary even more because $\bar{u}_{ii}$ of methane is practically independent of T. Similar results were obtained by Lal and Spencer (1973) by a Monte-Carlo simulation of the statistics of chain molecules for pure *n*-alkanes.[†] Another valid proof that the London energy of chain molecules depends on the temperature will be given in Chap. 6, in the discussion of equations of state.

Pitzer and coworkers (1955) introduced the *acentric factor* ω, which became the most popular empirical measure of the noncentral London energy. The acentric factor is defined as follows:

$$\omega = -\log(P/P^c)_{T/T^c} - 1.000 \tag{3.11}$$

where P/P^c is the value of the reduced vapor pressure at $T/T^c = 0.7$; therefore, it is a characteristic constant for a given fluid. Among several relations between the thermodynamic properties of fluids and ω, the authors developed a relation for $\beta(T)$ as a power series of ω. Although Kaul and Prausnitz (1977) have shown that the relation for

---

[†] The principles of the Monte-Carlo method are described by Metropolis et al. (1953).

$\beta(T)$ begins to fail for alkanes longer than *n*-heptane, it is possible that only the empirical relation fails, whereas the principles on which the concept of ω is based remain valid. This problem is considered again in Chap. 4. Reid and Sherwood (1958) have assembled the most complete collection of the values of ω and review some of the useful relations involving this parameter.

The temperature dependence of the other kinds of noncentral forces, those due to the dipole and the quadrupole moments, was recently confirmed by molecular dynamics "experiments" that consisted of a computer simulation of the motions and the interactions of molecules. In these studies Wang et al. (1974) and Haile (cited by Gubbins and Twu, 1977) demonstrated the strong effect of μ or Q on the residual internal energy of the fluid. These studies also led to useful formulas for the residual Helmholtz energy due to μ or Q as a function of the reduced temperature and the reduced density; these are considered in the next chapter.[†]

At very small intermolecular distances, when the electron clouds of two molecules begin to overlap, the attraction is replaced by a strong repulsion. Quantum mechanical calculations, outlined by Hirschfelder, Curtiss, and Bird (1954), suggest that the intermolecular repulsion is an exponential function of r, but more convenient forms are used in the theories of fluids. The function $u(r)$, including both repulsion and attraction, is shown schematically in Fig. 3.4. The subscripts *ij* were dropped in this figure and in the relations given further in this chapter.

Mie (1903) was the first to suggest that $u(r)$ might be expressed as the sum of two terms: a positive one varying with r^{-n} and a negative one varying with r^{-m}, where $n > m > 0$. *Mie's potential* in the reduced form is

$$u(r) = \frac{n\bar{u}}{n-m}\left(\frac{n}{m}\right)^{m/(n-m)}\left[\left(\frac{\sigma}{r}\right)^n - \left(\frac{\sigma}{r}\right)^m\right] \qquad (3.12)$$

where $\bar{u}$ is the minimum value of $u(r)$. Mie's potential with $m = 6$, suggested by London's theory, and $n = 12$, dictated by convenience, is known as the *Lennard-Jones potential*:

[†]Equations with numerical coefficients based on molecular dynamics simulations are treated as theoretical equations. Thus, molecular dynamics enjoys the unusual distinction of being simultaneously an experiment and a theory. Molecular dynamics and its applications are reviewed by Berne (1971) and by Berne and Forster (1971).

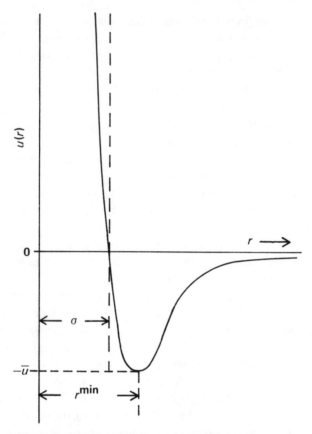

FIG. 3.4.   The variation of $u(r)$ from positive values at $r < \sigma$, due to a repulsion, to negative values at larger distances r between the centers of two molecules; σ is the collision diameter, and $\bar{u}$ is the minimum value of $u(r)$—by convention, a positive number.

$$u(r) = 4\,\bar{u}\left[\left(\frac{\sigma}{r}\right)^{12} - \left(\frac{\sigma}{r}\right)^{6}\right] \tag{3.13}$$

For this potential, the distance at which the minimum of $u(r)$ appears is $r^{\min} = 2^{1/6}\sigma = 1.1225\sigma$. By inserting a relation for $u(r)$ into (1.3), the values of $\bar{u}$ and σ can be determined from the experimental data for the second virial coefficient of a given substance. The values so obtained for the Lennard-Jones potential do not reproduce viscosity or other data for gases. Moreover, the values of $\bar{u}$ deviate very much, even for argon, from the more realistic values known today. The Lennard-Jones potential is used extensively in some theoretical calculations and in molecular dynamics simulations in which argon and krypton are used as the reference substances, and only relative numbers—not the true values of $\bar{u}_{oo}^L$ and σ_{oo} of the reference—are important. These results cannot be extended to large molecules because the potential then fails completely, and even the ratios $\bar{u}_{ii}^L/\bar{u}_{oo}^L$ are incorrect.

A much better approximation is the *Kihara model* (1953), according to which the molecules have hard cores of diameter $\sigma^h < \sigma$:

$$u(r) = \infty \quad \text{for} \quad r \leq \sigma^h;$$

$$u(r) = \frac{n\bar{u}}{n-m}\left(\frac{n}{m}\right)^{m/(n-m)}\left[\left(\frac{\sigma-\sigma^h}{r-\sigma^h}\right)^n - \left(\frac{\sigma-\sigma^h}{r-\sigma^h}\right)^m\right]$$

$$(r > \sigma^h) \tag{3.14}$$

For fixed n and m, there are three parameters in the Kihara potential: $\bar{u}$, σ, and σ^h. As pointed out by Sherwood and Prausnitz (1964), the degree to which three parameters can be unambiguously determined from the second virial coefficient data is strongly dependent on the accuracy and the temperature range of the experimental data.

The Kihara potential can be simplified to a two-parameter equation by putting σ^h/σ equal to a constant—for example, 0.88—and it remains superior to the Lennard-Jones potential. The superiority is due to the assumption of a hard core. As shown by Sherwood and Prausnitz (1964), the *square-well potential* is as good as the Kihara potential in spite of a completely different function for $u(r)$—retaining, however, the hard core. This potential is expressed as

$$u(r) = \infty \qquad \text{for} \quad r \leq \sigma;$$

$$u(r) = -\bar{u} \quad \text{for} \quad \sigma < r \leq \mathcal{R}\sigma;$$

$$u(r) = 0 \qquad \text{for} \quad r > \mathcal{R}\sigma \tag{3.15}$$

where $(\mathcal{R} - 1)$ is the width of the well in r/σ units as shown in Fig. 3.5. The square-well potential was introduced by Kirkwood (1935), who called it the *step function*. The values of $\mathcal{R}$ obtained by Sherwood and Prausnitz from $\beta(T)$ data for several substances appear to be

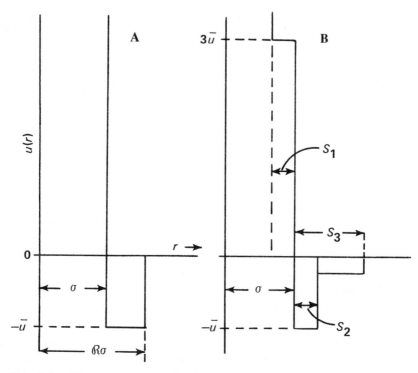

FIG. 3.5.   The square-well potential (A) and the three-step potential (B). The height of the first step, $3\bar{u}$, and the depth of the third step, $0.21\ \bar{u}$, were established as the best choices for argon and methane.

nearly constant, $\mathfrak{R} = 3/2$. Thus, the thickness of the "shell of interactions" amounts to about $1/2\ \sigma$.

Also in the *three-step potential*, proposed by Kreglewski (1970)[†] and shown in Fig. 3.5, the ratios s_1/σ and s_2/σ appear to be approximately constant. Chen and Kreglewski (1977) found by applying an equation of state to *PVT* data that

$$s_1/\sigma \approx 0.12 \tag{3.16}$$

for most fluids without hydrogen bonds (bridges). At least for argon and methane, the height of the first step is approximately equal to $3\bar{u}$. Integration according to (1.3) for this potential yields

[†]Unpublished work. The three-step potential is used in the Thermodynamic Research Center for smoothing second virial coefficient data.

$$\frac{\beta(T)}{2/3 \ (\pi N_o \sigma^3)} = \mathcal{R}_1 - (\mathcal{R}_1 - 1)(e^{-3\bar{u}/kT} - 1)$$
$$- (\mathcal{R}_2 - 1)(e^{+\bar{u}/kT} - 1)$$
$$- (\mathcal{R}_3 - \mathcal{R}_2)(e^{+0.21\bar{u}/kT} - 1) \qquad (3.17)$$

where

$$\mathcal{R}_1 = \left(1 - \frac{s_1}{\sigma}\right)^3 ; \ \mathcal{R}_2 = \left(1 + \frac{s_2}{\sigma}\right)^3 ; \ \mathcal{R}_3 = \left(1 + \frac{s_3}{\sigma}\right)^3$$

It simplifies to the square-well potential when $\mathcal{R}_1 = 1$ and $\mathcal{R}_3 - \mathcal{R}_2 = 0$. The third step in this potential is without any significance in the theory of fluids. It merely improves the agreement with experimental $\beta(T)$ data. The first step, however, accounts approximately for the temperature dependence of σ, which should not be neglected. This problem is discussed in Chap. 6. Kincaid, Stell, and Goldmark (1976) applied a similar potential with an identical first step. They found it sufficient to account for melting and solid-phase transition phenomena. For a small range of s_1 values, the coexistence curves of the fluid-solid and the solid-solid transition meet at a triple point. Moreover, they found that a theoretical fluid having the first-step potential ($s_1 \neq 0$) but no attraction ($s_2 = s_3 = 0$) exhibits a gas-liquid critical point ! .

These results show that we may describe the physical properties of a fluid by using a crude approximation to the shape of $u(r)$, such as a *two-step potential*, with only the first and the second steps. Integrations involving this potential are greatly simplified compared with the Mie's or Kihara potential, not to mention more elaborate forms. Most important, however, is that with this model we may apply directly the relations for the Helmholtz energy of hard spheres considered in Chap. 6. The square-well potential, implying hard spheres under all conditions, is oversimplified. It does not take into account the variation of σ with T, which can be interpreted as being due to overlapping or deformation of the electron clouds upon collisions at high temperatures, nor does it lead to the transition phenomena considered by Kincaid and his colleagues (1976).

More elaborate potential forms are reviewed by Fitts (1966), Barker and coworkers (1974), and Lee, Neufeld, and Bigeleisen (1977).

Weeks, Chandler, and Andersen (1971) introduced a new potential composed of two parts, $u^{\text{rep}}(r)$ for repulsion and $u^{\text{at}}(r)$ for attraction, the sum of which is equal to the Lennard-Jones potential $u^{\text{LJ}}(r)$, given by (3.13):

$$u^{\text{rep}}(r) = u^{\text{LJ}}(r) + \bar{u}^{\text{LJ}} \quad \text{for} \quad r < 2^{1/6}\sigma$$

$$\qquad\qquad = 0 \qquad\qquad\quad \text{for} \quad r \geq 2^{1/6}\sigma;$$

$$u^{\text{at}}(r) = -\bar{u}^{\text{LJ}} \qquad \text{for} \quad r < 2^{1/6}\sigma$$

$$\qquad\quad = u^{\text{LJ}}(r) \qquad\quad \text{for} \quad r \geq 2^{1/6}\sigma \qquad (3.18)$$

so that $u^{\text{rep}}(r) + u^{\text{at}}(r) = u^{\text{LJ}}(r)$. Here, $\bar{u}^{\text{LJ}}$ is the minimum value of $u^{\text{LJ}}(r)$. This unusual potential was chosen to evaluate the effective collision diameters which appear to be functions of both temperature and density. The repulsion $u^{\text{rep}}(r)$, given by (3.18), is much harder than that varying with $(\sigma/r)^{12}$. One of the important results is that if the fluid density is sufficiently large, $g(r)$ is accurately approximated by $g^{\text{rep}}(r)$, where $g^{\text{rep}}(r)$ is a function of $u^{\text{rep}}(r)$ only. In other words, as stated earlier by Longuet-Higgins and Widom (1964), "at high densities the structure of a liquid is mainly determined by the repulsion forces."

These statements, as well as Kincaid's results, mentioned above, may create an impression that the forces of attraction are unimportant in the theory of fluids. Although the structure and $g(r)$ of a dense fluid depend mostly on $g^{\text{rep}}(r)$, the thermodynamic properties depend on both $g(r)$ and *total* $u(r)$.

It is customary to assume that the various parts of $u(r)$ are additive; that is,

$$u(r) = u^{\text{rep}}(r) + u^{L}(r)$$

$$\qquad\quad + u^{\mu\mu}(r) + u^{\mu Q}(r) + u^{QQ}(r) + \ldots \qquad (3.19)$$

where $u^{\text{rep}}(r)$ is positive and the remaining terms are negative. The London energy $u^{L}(r)$ may include its noncentral, temperature-dependent part for nonspherical molecules. Assuming that this part varies with r in the same way as the "central" part, following Rowlinson (1969), we may write for the London energy

$$\bar{u}^{L} = \bar{u}^{o}\left(1 + \frac{\eta^{L}}{kT}\right) \qquad (3.20)$$

and

$$u^L(r) = u^o(r)\left(1 + \frac{\eta^L}{kT}\right) \quad \text{for} \quad r > \sigma$$

$$u^L(r) = \infty \qquad\qquad\qquad \text{for} \quad r < \sigma \qquad\qquad (3.21)$$

where η^L is positive.[†] It should be remembered that London's formula is an approximate relation and that there are higher, r^{-8} and r^{-10}, terms (Hirschfelder, Curtiss, and Bird, 1954). The additivity assumed in (3.19) is probably a good approximation for small, nearly spherical molecules whose $u^L(r)$ is (nearly) independent of orientation. Hydrogen chloride—for which all the terms given in (3.19) are non-zero—nitrogen, and oxygen with weak $u^{QQ}(r)$ in addition to $u^L(r)$ are examples of such molecules.

The various parts of $u(r)$ may also be additive in *n*-alkyl derivatives, $C_nH_{2n+1}X$, where the C-X bond is a source of polarity. It is possible here that the mutual orientation of the molecules favored by $u^L(r)$ coincides with that favored by $u^{\mu\mu}(r)$. The critical temperatures T^c of $C_nH_{2n+1}X$ are always higher than those of the *n*-alkanes with the same number of carbon atoms. The temperature T^c is not a direct measure of $u(r)$ for liquids with noncentral forces, but the ratios T^c_{AX}/T^c_A for alkyl derivatives/alkanes are instructive in qualitative considerations. These ratios for various series of *n*-alkyl compounds always decrease and tend toward unity with the increasing length of the chain. This phenomenon is easily explained by the theory of noncentral forces. Following Rowlinson (1969), we can write, instead of the Lennard-Jones potential,

$$u(r) = 4\,\bar{u}^o\left\{\left(\frac{\sigma}{r}\right)^{12} - \left(\frac{\sigma}{r}\right)^6[1 + 2\chi]\right\} \qquad (3.22)$$

or, more generally,

$$u(r) = u^{\text{rep}}(r) - 4\,\bar{u}^o\left(\frac{\sigma}{r}\right)^6(1 + 2\chi) \qquad (3.23)$$

[†]Eq. (3.21) has been extensively tested by calculations of U^r/RT of alkanes by means of an accurate equation of state (Chap. 6). Heintz and Lichtenthaler (1977) proposed a different temperature dependence of U^r for mixtures of chain molecules. It has not yet been tested for pure fluids or compared with the molecular beam scattering data given in Table 3.1.

where in the case of $\mu\mu$ interactions, from (3.8),

$$\chi^{\mu\mu} = \mu_i^2\mu_j^2/(24\ \bar{u}^o\ \sigma^6\ kT) \tag{3.24}$$

If the various parts of noncentral energy contribute separately and are additive, then

$$u(r) = u^{\text{rep}}(r) - 4\ \bar{u}^o\left(\frac{\sigma}{r}\right)^6[1 + 2(\chi^L + \chi^{\mu\mu})] \tag{3.25}$$

or

$$u(r) = u^{\text{rep}}(r) + u^o(r)\left[1 + \frac{(\eta^L + \eta^{\mu\mu})}{kT}\right] \tag{3.26}$$

where

$$\eta^{\mu\mu}/kT = u^{\mu\mu}(r)/u^o(r) \tag{3.27}$$

It will be shown later that both u^o and η^L increase with the chain length of the molecules. Hence, the relative contribution of $X^{\mu\mu}$ or $\eta^{\mu\mu}$ decreases rapidly even if the dipole moment (μ) slowly increases—as, for example, in the series CH_3Cl (1.86 D), C_2H_5Cl (2.01 D), and n-C_3H_7Cl (2.04 D), where 1 D (Debye) $= 10^{-18}$ e.s.u. (Le Fevre, 1948). Meyer and his colleagues (1966, 1971, 1976) estimated the relative contributions of London energies and dipolar interactions to the enthalpy of vaporization of a large number of organic liquids. The latter rapidly decrease with the increasing size of the homologues.

In the case of perfluoro-n-alkanes, the situation is different. They are nonpolar but have relatively strong multipole moments (see, for example, Parsonage and Scott, 1962, and Stogryn and Stogryn, 1966). The quadrupoles are formed along the F-C-F bond lines, perpendicular to the axis along the molecule. If $u^L(r)$ is the dominating energy and the molecules align themselves favorably for $u^L(r)$ (Fig. 3.1a), the configuration does not necessarily enhance $u^{QQ}(r)$ (Fig. 3.3b). If the orientations are not favorable for $u^{QQ}(r)$, it will be weaker than the statistical average calculated in the absence of $u^L(r)$. The ratios of T^c of perfluoro-n-alkanes to n-alkanes are as follows:

Number of C atoms	1	2	3	4
T^c_{FA}/T^c_A	1.194	0.9590	0.9332	0.9088

Number of C atoms	5	6	7	8
T^c_{FA}/T^c_A	0.898	0.8822	0.8788	0.883

The ratio is quite large for CF_4, and it is probably due mostly to octopole interactions. These interactions do not suffice to raise T^c of the next homologues above those of *n*-alkanes. It seems that $u^L(r)$ and $u^{multipole}(r)$ are not additive in this case, but competitive.

The contributions of the various terms in (3.19) depend on the chemical nature of the substance and thus, obviously, on the nature of the bonds between the atoms. In studying the thermodynamic properties of pure fluids and mixtures—particularly the second virial coefficients of gases, the residual free energy of liquids and liquid mixtures, and their *temperature dependence*, compared with the same properties of alkanes or cycloalkanes—we note that some substances appear to be "abnormal." Their abnormality manifests itself in the values of η/k and, particularly, in $\bar{u}_{ij}$ of mixtures. These compounds exhibit one common feature: the *electronegativities of two or more elements*, forming one or more bonds in the molecule, *are very different*. The electronegativity values E of some elements, given by Pauling (1967), and the atomic radii, selected by Moelwyn-Hughes (1957), are collected in Table 3.2. The values of E, calculated recently by Batsanov (1975), differ only slightly from these values.

According to Pauling, the amount of ionic character of a *single* bond *A-B* formed by elements *A* and *B* is

$$I_{AB} = 1 - \exp\left[-\frac{1}{4}(E_A - E_B)^2\right] \tag{3.28}$$

TABLE 3.2.   The values of electronegativity and the atomic radii of some elements.

	E	$r_A/\text{Å}$		E	$r_A/\text{Å}$
H	2.1	0.375	S	2.5	1.04
C	2.5	0.771	F	4.0	0.68
Si	1.8	1.174	Cl	3.0	(1.02)
N	3.0	0.70	Br	2.8	1.19
P	2.1	1.10	I	2.5	1.35
O	3.5	0.60			

The resulting *bond dipole moment* or *bond moment* is, to a first approximation, proportional to $(E_A - E_B)$. However, as shown by Mulliken (1935), there are two additional terms: the induced moment and a term depending on the difference in the size of the atoms that may either assist or oppose the main dipole term. The size term (also called the homopolar dipole term) may exist even when $E_A \approx E_B$ and may be fairly large. Its sign is such that *its positive pole is always directed toward the larger atom* (Mulliken, 1935). For example, for the C-H bond and for the Si-H bond, I_{AB} equals 0.04 (or 4 percent) and 0.02 (2 percent), respectively. By considering, however qualitatively, the size term, we obtain the following schemes:

Electronegativity term	$(-)$	$(+)$		$(+)$	$(-)$
Bond	C	— H		Si	— H
Size difference term	$(+)$	$(-)$		$(+)$	$(-)$

The resulting bond moment may be approximately equal to zero in the first case but not in the second case. Therefore, alkanes and cycloalkanes may be expected to interact through London energy only. In fact, methane closely follows the principles established for the rare gases (see Chap. 4), and there are no examples that would prove that alkanes may interact in some specific manner with other organic substances, or form complexes in the liquid or in the gas phase. They are widely used as solvents in spectroscopic investigations of electron donor + acceptor complexes. Like argon and krypton, alkanes and cycloalkanes will be further called *inert solvents*.

The foregoing scheme also implies that CS_2 is a nearly inert solvent, but CCl_4 certainly is not. When $(E_A - E_B)$ is large, an "abnormal" behavior can be expected. For example, CO_2 has large Q, SO_2 has both large μ and large Q, perfluoro-alkanes have large multipole moments. Ketones are more "associated" than are amines.

The methods of calculation of dipole moments of molecules from bond and group moments (and vice versa) are critically reviewed by Minkin, Osipov, and Zdanov (1970). The most extensive tables of experimental dipole moments are due to McClellan (1963) and J. W. Smith (1955).

The quadrupole moments, selected by Stogryn and Stogryn (1966) and others, are presented in Table 3.3. Here, not only the values but also the signs are important. The signs for the last five com-

TABLE 3.3.   The values of quadrupole moments of some molecules.

Molecule	$Q \cdot 10^{26}/$ e.s.u.cm^2
Hydrogen	+0.662
Hydrogen deuteride	+0.642
Deuterium	+0.649; +0.643[a]
Nitrogen	−1.52; −1.4[a]
Oxygen	−0.39; −0.4[a]
Fluorine	+0.88
Chlorine	+6.14
Nitrogen oxide	(−1.8); −4.2[a]
Dinitrogen oxide	(−3.0); −4.2[a]
Ammonia	−1.
Carbon monoxide	−2.5
Carbon dioxide	−4.3; −4.3[a]
Carbonyl sulfide	−2.0[a]
Carbon disulfide	+4.26[b]
Hydrogen fluoride	+2.6
Hydrogen chloride	+3.8
Hydrogen bromide	+4.
Hydrogen iodide	+6.
Acetylene	+5.10[c]
Ethylene	(+1.5); +3.85[c]
Ethane	−0.65
Benzene	(−14.5);[d] −9.98[b]
Napthalene	−13.5[e]
Perfluorobenzene	+9.5[b]
Sulfur dioxide	4.4
Hydrogen cyanide	4.4
Chlorine cyanide	6.6
Bromine cyanide	6.8
Dichloromethane	4.1

SOURCES: (a) Buckingham (1970); (b) Ritchie and Vrbancich (1980); (c) Spurling and Mason (1967); (d) Hanna (1968); (e) Calvert and Ritchie (1980).

pounds listed are not known. These data show that the schemes given above, suitable for single bonds, are incomplete. Even the molecules containing identical atoms but double or triple bonds may not be inert, for example, H_2, N_2, and O_2. The $u^{QQ}(r)$ of nitrogen is large compared to $u^L(r)$. It is surprising that the thermodynamic properties of

TABLE 3.4.   The values of $\bar{u}/k$ in Kelvins obtained by several methods.

	Method	Authors
Argon	$\beta(T)$ data and	
147.2	Kihara potential	Sherwood and Prausnitz (1964)
166	" "	Pope et al. (1973)
163.7	" "	Weir et al. (1967)
147.5	Thermal neutron scattering by liquid	Brostow (1975)
146.3	X-ray diffraction by dense fluid	Karnicky et al. (1976)
142.1	$\beta(T)$, viscosity, solid *PVT*	Barker et al. (1974)
Krypton	$\beta(T)$ data and	
215.6	Kihara potential	Sherwood and Prausnitz (1964)
213.9	" "	Weir et al. (1967)
201.9	$\beta(T)$, viscosity, solid *PVT*	Barker et al. (1974)
Hydrogen		
(46.0)[a]	$\beta(T)$ and Kihara potential	Prausnitz and Myers (1963)
35.95[a]	Viscosity and Gegenbach-Hahn-Schrader-Toennies potential	Clifford et al. (1975)

(a) Including quantum corrections at low temperatures.

pure N_2 and of the Ar + N_2 system, analyzed later, are so close to those of an inert solvent. The effect of Q in C_2H_6 must be entirely negligible, whereas C_2H_2 is a very "abnormal" compound. Also, the olefins and aromatic hydrocarbons are capable of donor + acceptor interactions (Chap. 9).

In concluding this chapter, it is interesting to compare the values of $\bar{u}$ determined by various methods for some of the small molecules. For the larger globular or chain molecules, the values of $\bar{u}$ were usually assumed to be independent of temperature, so they are only average values in a certain temperature range.

The values of $\bar{u}$ for argon, krypton, and hydrogen are given in Table 3.4. The uncertainty of the best values is estimated by Barker and coworkers (1974) to be ± 5 percent. Probably the most accurate value for argon is that due to Brostow (1975). They appear to be very close to the critical temperatures: 150.86 K for argon, 209.4 K for krypton, and 33.2 K for hydrogen. The results obtained by Chen and Kreglewski (1977), given in Chap. 6, show that this is true for all small spherical molecules.

The values of u/kT^c for the Lennard-Jones potential, obtained by Sherwood and Prausnitz (1964) from $\beta(T)$ data for Ar, Kr, Xe, and CH_4, are 0.780, 0.783, 0.767, and 0.781, respectively, the most probable value being

$$\bar{u}^{LJ}/kT^c \approx 0.780 \tag{3.29}$$

The values of u/kT^c for the square-well potential with a constant width $\mathcal{R} = 3/2$, obtained by Beret and Prausnitz (1975) by means of an accurate equation of state, are

$$\bar{u}^{SW}/kT^c = 0.7928 \tag{3.30}$$

for argon and nearly the same values for other small molecules.

CHAPTER 4

Deviations from the Principle of Corresponding States

The statistical-mechanical theory of liquids and liquid mixtures provides relations between the thermodynamic functions of the system and the interaction energies $u(r)$, which involve distribution functions for sets of molecules: for example, pairs, triplets, and so on, which form subsets of the total number N of molecules in the system.

The pioneering experimental studies of radial (pair) distribution functions in liquids were performed by Debye and Menke and by Hildebrand and his coworkers in the 1930s. They are described by Hildebrand and Scott (1950, 1962). The fundamental integro-differential equations for the distribution functions were formulated by Kirkwood, Born and Green, and Mayer, and by Kirkwood and Salsburg (1953), who review the other works mentioned. These equations could not be solved to obtain practical results until the more recent development of the theory of hard-sphere fluids.

As shown in Chap. 1, the knowledge of $A(T, V, x_i, x_j, \ldots)$ and $Z(T, V, x_i, x_j, \ldots)$ derived from it (or vice versa) suffices to perform all the phase equilibrium calculations for pure fluids or mixtures. The basic relation for the Helmholtz energy of pure fluids and mixtures has been obtained from the perturbation theory of Barker and Henderson (1967, 1968), Henderson and Barker (1968), and Leonard, Henderson, and Barker (1970). If the small term due to the difference $(d_{12} - \delta_{12})$—effective collision diameters σ_{12}^c calculated from two different relations—is neglected, the relation for a binary system is

$$\frac{A^r}{NkT} = \frac{A^{rh}}{NkT} + \sum_i \sum_j x_i x_j \left\{ \frac{2\pi\rho_n}{kT} \int_\sigma^\infty u_{ij}(r) g_{ij}^h(r) \ r^2 \ dr \right.$$

$$\left. - \frac{\pi\rho_n}{kT} \left(\frac{\partial\rho_n}{\partial P} \right)_T^{(h)} \int_\sigma^\infty u_{ij}^2(r) g_{ij}^h(r) \ r^2 \ dr \right\} \tag{4.1}$$

where $\rho_n = N/V$; A^{rh}, $g_{ij}^h(r)$, and $(\partial\rho_n/\partial P)_T^{(h)}$ are the residual Helmholtz energy, the radial distribution function, and the compressibility, respectively, of a binary *mixture of hard spheres*. The function A^{rh} and the other functions for hard spheres are known and are given in Chap. 6. Analytical expressions for $g_{ij}^h(r)$ of pure fluids are given by Watts and Henderson (1969), and of mixtures by Lebowitz (1964).

The last term in (4.1) is much smaller than the second one; thus any higher terms with pair interactions must be entirely negligible. This is one of the essential results of this theory because the second, or main, term was already known. The three-body term, not given in (4.1), will be considered in Chap. 8 in connection with solubilities of gases in liquids.

Rogers and Prausnitz (1971) applied this theory to the calculation of vapor-liquid equilibria up to the critical points of the argon + neopentane and the methane + neopentane systems; they obtained an excellent agreement with the experimental results. However, the very lengthy computations required for obtaining $g_{ij}^h(r)$ of mixtures tend to exclude this method from practical calculations at the present time.

This example shows that the theory of fluids has made great progress in recent years. Other approaches to the problem are reviewed by Barker and Henderson (1976) and by Boublik (1977). We shall consider only the methods in which the problem of evaluating $g_{ij}(r)$ of mixtures is either bypassed completely by the use of the principle of corresponding states, or $g_{ij}(r)$ is expressed by a relatively simple function of the reduced density ρ^* and temperature T^*. Although it is customary to write $g(r)$, in fact it is $g(r, \rho^*, T^*)$.

The idea of corresponding states of fluids arose from the van der Waals equation of state and was suggested by him 110 years ago. The principle relates the *PVT* properties of one pure substance to those of another as follows:

$$P_i(T, V) = \left(\frac{T_i^c V_o^c}{T_o^c V_i^c} \right) P_o \left(\frac{T T_o^c}{T_i^c}, \frac{V V_o^c}{V_i^c} \right) \tag{4.2}$$

where subscript o denotes a reference substance. If argon is chosen as the reference substance, then (4.2) appears to be valid only for other small spherical molecules such as krypton or methane. The validity was demonstrated by Pitzer (1939) and by Guggenheim (1967).[†]

The principle of corresponding states, derived by Pitzer (1939)

[†] For deviations from the principle due to small molecular mass and very low temperatures, and the corrections derived from quantum mechanics for such fluids (He, Ne, H_2), see Hirschfelder, Curtiss, and Bird (1954).

by means of statistical mechanics, has a broader basis and should be valid for a larger group of substances. The basic assumptions are discussed by Rowlinson (1969), and his treatment is followed here. First, the configurational energy $\mathcal{U}$ of N mixed molecules of species i and j is a sum of pair interactions:

$$\mathcal{U} = \sum_{\alpha < \beta} \sum u_{ij}(r) \tag{4.3}$$

where α and β denote each two of the N molecules.[†] Triple interactions,

$$\sum_{\alpha} \sum_{< \beta <} \sum_{\gamma} u_{ijk}(r),$$

which should be added to (4.3), are neglected. There is some indirect evidence that u_{ijk} contributes 5 to 10 percent of the total energy of liquid argon at its triple point and that this contribution is positive; that is, it weakens the interaction energy. As pointed out by Barker and his colleagues (1974), the three-body effects are appreciable, but the deviations from the principle of corresponding states depend on these effects divided by $u_{ij}\sigma_{ij}^9$, so they are much smaller than 10 percent. This statement is applicable to nonpolar molecules only. The three-body effects in mixtures become significant at infinite dilutions—on the solubility of gases at very low pressures, for example (Goldman, 1978), as will be demonstrated in Chap. 8.

Secondly, for two substances having $u(r) = 0$ at $r = \sigma$ and

$$u_{11}(r) = \bar{u}_{11}F(\sigma_{11}/r)$$
$$u_{22}(r) = \bar{u}_{22}F(\sigma_{22}/r), \tag{4.4}$$

$F(\sigma_{ii}/r)$ is assumed to be a common function. Eq. (4.4) can also be expressed in the more general form

$$u_{ij}(r) = f_{ij}\bar{u}_{oo}(r\sigma_{oo}/\sigma_{ij}) \tag{4.5}$$

where subscript oo denotes a reference substance and $f_{ij} = \bar{u}_{ij}/\bar{u}_{oo}$. Longuet-Higgins (1951) called fluids whose $u(r)$ have a common function $F(\sigma/r)$, or conform to the same form of the potential, *conformal fluids* or *conformal solutions* (mixtures). If (4.5) is limited at the pres-

---

[†] For example, if there are three molecules, $N = 3$, α is 1 or 2, β is 2 or 3, and the summation is carried out for $1 + 2$, $2 + 3$, and $1 + 3$ pairs.

ent time to pure fluids, one obtains from the assumptions (4.3) and (4.5) that the configurational partition function of the fluid i is

$$Q_i(T, V) = h_{ii} Q_o(T/f_{ii}, V/h_{ii})$$

where $h_{ii} = (\sigma_{ii}/\sigma_{oo})^3$. By applying the basic relation (1.8), we obtain the configurational Helmholtz energy:

$$A_i(T, V) = f_{ii} A_o(T/f_{ii}, V/h_{ii}) - 3NkT \ln(\sigma_{ii}/\sigma_{oo}) \qquad (4.6)$$

Differentiation of (4.6) with respect to the volume gives, for a homogeneous fluid,

$$P_i(T, V) = (f_{ii}/h_{ii})P_o(T/f_{ii}, V/h_{ii}) \qquad (4.7)$$

Eqs. (4.6) and (4.7) are expressions of the principle of corresponding states. If the state of the reference substance is represented by a P, V, T surface, then the states of all conformal substances will be geometrically identical with it when the scales of P, V, and T are multiplied by (f_{ii}/h_{ii}), h_{ii}, and f_{ii}, respectively. Equations of state expressed in terms of these quantities are called *reduced equations of state*, and quantities such as $kT/\bar{u}$, V/σ^3, and $\rho\sigma^3$ are the reduced temperature, volume, and density, respectively.

Eq. (4.2) is a special case of (4.7) when $f_{ii} = T_i^c/T_o^c$ and $h_{ii} = V_i^c/V_o^c$. This is true only for small spherical molecules. For nonspherical molecules, the close-packed volume of a fluid or the effective diameter based on the radius of gyration become better measures of the *average* σ than does the critical volume V^c; also, f_{ii} is certainly not equal to T_i^c/T_o^c and is a function of the temperature (and of the reduced density, for large molecules). The principle expressed by (4.7) is much more general than the empirical principle expressed by (4.2), and it will be shown later that the former is valid also for large molecules with London energies only. Even in the presence of a strong quadrupole moment, the deviations from the principle are relatively small. Benzene conforms to an equation of state with the universal constants fitted to the data for argon nearly as well as does *n*-hexane. Hydrogen chloride, to the contrary, does not conform to the principle, in spite of the fact that the μ-μ interactions vary with r^{-6} and are thus conformal with London interactions as expressed by (3.23) and (3.24). As shown by Stell, Rasaiah, and Narang (1972), the reason is that the three-body interactions $u_{ijk}^{\mu\mu}(r)$ cannot be neglected. Nevertheless, the principle remains a very useful tool because it appears that the deviations due to the presence of dipoles or multipoles can again be expressed as universal functions of the reduced density and the reduced temperature.

A. Approximations for Mixtures

The theory of mixtures of molecules that satisfy (4.2) and (4.6) is a difficult problem. Longuet-Higgins (1951) derived a theory of conformal mixtures that is exact to the first order in the differences $(f_{ij} - 1)$ and $(\sigma_{ij}/\sigma_{oo}) - 1$. A rigorous derivation of the theorem for mixtures is due to Salsburg, Wojtowicz, and Kirkwood (1957) and Wojtowicz, Salsburg, and Kirkwood (1957).

Extensions of the theory of conformal mixtures to systems in which f_{ij} and σ_{ij}/σ_{oo} differ greatly from unity invariably led to completely erroneous results for the excess functions and other properties. Common to all these theories has been the assumption either that the distribution of all the species in a mixture is random or that the deviations from randomness are sufficiently small to permit the application of some perturbation technique. These theories are reviewed by Leland, Rowlinson, and Sather (1968). Since the theories fail even for small spherical molecules, as in the $CH_4 + CF_4$ system, the reasons must be other than non-randomness.

The theories were invariably based on the Mie's potential (3.12). For this potential, $F(\sigma/r)$ in (4.4) is proportional to $(\sigma/r)^n - (\sigma/r)^m$. The *average potential energy* of a random mixture (Prigogine et al., 1957; Byers Brown, 1957; R. L. Scott, 1956) is

$$f_x F(\sigma_x/r) = x_i^2 f_{ii} F(\sigma_{ii}/r) + 2x_i x_j f_{ij} F(\sigma_{ij}/r) + x_j^2 f_{jj} F(\sigma_{jj}/r) \quad (4.8)$$

If n and m are the same for all the components,[†] $F(\sigma/r)$ can be separated; that is, $f_x(\sigma_x/r)^n = x_i^2 f_{ii}(\sigma_{ii}/r)^n + 2_i x_j f_{ij}(\sigma_{ij}/r)^n + x_j^2 f_{jj}(\sigma_{jj}/r)^n$, and similarly for the attraction terms $(\sigma/r)^m$. From the two relations, we deduce

$$f_x = \left[\sum_i \sum_j x_i x_j f_{ij}(\sigma_{ij}/\sigma_{oo})^m \right]^{\frac{n}{n-m}} \left[\sum_i \sum_j x_i x_j f_{ij}(\sigma_{ij}/\sigma_{oo})^n \right]^{\frac{-m}{n-m}} \quad (4.9)$$

[†] Second virial coefficients of mixtures in which the components have different values of n and m are considered by Schafer (1977).

and

$$\sigma_x/\sigma_{oo} = \left[\sum_i \sum_j x_i x_j f_{ij}(\sigma_{ij}/\sigma_{oo})^m\right]^{\frac{-1}{n-m}}$$

$$\times \left[\sum_i \sum_j x_i x_j f_{ij}(\sigma_{ij}/\sigma_{oo})^n\right]^{\frac{1}{n-m}} \tag{4.10}$$

Eqs. (4.9) and (4.10) are the basic relations of the so-called *random mixing approximation*. Other and better relations are possible for random mixtures. To avoid any confusion, we shall call the above relations the *Mie's approximation* or the *Lennard-Jones (LJ) approximation* in the case of $n = 12$ and $m = 6$.

The van der Waals approximation (1890), following from the extension of his equation of state to mixtures, in which the constant a is proportional to $\bar{u}\sigma^3$ and constant b to σ^3 (see, for example, Hill, 1960), is

$$f_x h_x = \sum_i \sum_j x_i x_j f_{ij} h_{ij}; \tag{4.11}$$

$$h_x = \sum_i \sum_j x_i x_j h_{ij} \tag{4.12}$$

The plausible form of $u(r)$ corresponding to this approximation is considered later. For a binary system,

$$h_x = x_1^2 h_{11} + 2x_1 x_2 h_{12} + x_2^2 h_{22}$$

$$= x_1 h_{11} + x_2 h_{22} + x_1 x_2 (2h_{12} - h_{11} - h_{22})$$

According to van der Waals,

$$h_{ij} = (h_{ii} + h_{jj})/2 \tag{4.13}$$

and therefore,

$$h_x = \sum_i x_i h_{ii} \tag{4.14}$$

It should be emphasized that the van der Waals approximation is not tied to his equation of state. This approximation has been combined with much more accurate equations of state for mixtures of hard spheres to evaluate the excess functions of liquid mixtures. The results so obtained by Leland, Rowlinson, and Sather (1968) for the CH_4 + CF_4 system are compared below with the LJ approximation. The parameters obtained from very accurate values of $\beta(T)$, determined by Douslin, Harrison, and Moore (1967), are as follows (CH_4 is the component 1 and the reference substance):

Interaction	f	σ/σ_{oo}
1 + 1	1.000	1.000
1 + 2	0.917	1.132
2 + 2	1.017	1.242

The excess Gibbs energy and the excess volume of the liquid mixture at $x = 0.5$, calculated by using the above parameters, are as follows:

	$G^E/J \cdot mol^{-1}$	$V^E/cm^3 mol^{-1}$
Experiment[†]	360	0.88
LJ approximation	1260	13.9
VDW approximation	280	0.90

Thus, the Lennard-Jones approximation leads to very large errors, although $(\sigma_{22}/\sigma_{11}) - 1$ is small, 0.242, because it overstates the effect of the differences in molecular size. The van der Waals approximation is supposed to fail in the case of size differences larger than those in the foregoing system (Ewing and Marsh, 1977), but these tests are not conclusive, because the temperature dependence of $\bar{u}/k$ was neglected.

Stecki and Gradzki (1971) extended the Mie's and van der Waals approximations to polymer mixtures. The relations derived have never been tested for mixtures of molecules of very different sizes nor have they been applied in connection with an equation of state (to take care of the repulsion). Mansoori (1972) and Mansoori and Ali (1974) developed a new variational approach to mixtures. Subsequently, Mansoori and Leland (1972) and Lan and Mansoori (1975) extended the van der Waals model to include three-body interactions. The calculated results for high-pressure liquid-vapor equilibria are not better than those Rogers and Prausnitz (1971) obtained by using (4.1) where the three-body interactions are ignored. It appears that Barker's statement, mentioned in the discussion of (4.2), is also true for mixtures.

[†]Thorp and Scott (1956); Croll and Scott (1958).

For the square-well (SW) potential with a constant $\mathcal{R}$, $F(\sigma/r)$ is identical in all the terms of (4.8), so that

$$f_x = \sum_i \sum_j x_i x_j f_{ij} \tag{4.15}$$

and h_x is given by (4.14), because it results from the hard repulsion characteristic for this potential. As f_x is here independent of σ, Eq. (4.15) may be called the *square-well (O) approximation*, to distinguish it from another possible approximation involving the SW potential.

For the SW potential, $F(\sigma/r)$ is not exactly constant; it seems to decrease very slowly with the increasing size of the molecules. Recently, Kaul and Prausnitz (1977) proposed an SW potential with a well width

$$\mathcal{R} = 1 + 2/\sigma \qquad (\sigma \text{ in angstrom units, Å}) \tag{4.16}$$

The collision diameters of chain molecules σ_{jj} were calculated from the radius of gyration R defined by

$$R^2 = \frac{(\text{product of } q \text{ principal moments of inertia})^{1/q}}{\text{mass}} \tag{4.17}$$

where $q = 2$ for a linear molecule, and $q = 3$ otherwise. The relation is

$$\sigma_{jj} = \sigma_{ii} + 2(R_j - R_i) \tag{4.18}$$

where σ_{ii} and R_i are the values for the methane molecule, 3.350 Å and 0.44 Å, respectively. For a chain molecule, R_j represents the average of many possible configurations. A Monte-Carlo computer study by Bellemans (1973) shows that for a chain molecule,

$$(R_j)^2 = 0.164 \, m_c^{1.2}(1 + 0.547/m_c) \tag{4.19}$$

where m_c is the number of chain linkages.

With $\mathcal{R}$ fixed by (4.18) and (4.19), only the value of $\bar{u}$ in the potential remains to be found from the experimental data. The agreement obtained with the $\beta(T)$ data for the system with the most different sizes of the components for which data exist, $CH_4 + n\text{-}C_{10}H_{22}$, is excellent. We shall call the approximation with $\mathcal{R}$ given by (4.16) *the square-well (−1) approximation*, since it varies with σ^{-1}.

Kac, Uhlenbeck, and Hemmer (1963) derived the VDW attraction term exactly as the limit for $u(r)$ that is very weak and for which the intermolecular force $du(r)/dr = 0$. The VDW term results also from the following, more realistic conditions. The radial distribution function $g^h(r)$ of hard spheres has a sharp peak near $r \geq \sigma$ and is close to unity for larger distances r. If this curve is expressed by a one-step function

$$g^h(r) = \text{constant} = \bar{g}, \qquad \text{for } \sigma \leq r \leq \mathcal{R}_2\sigma$$

$$g^h(r) = 1, \qquad\qquad \text{for } r > \mathcal{R}_2\sigma \tag{4.20}$$

and $u(r)$ by the SW potential with $u(r) = -\bar{u}$ in the same range of r, from σ to $\mathcal{R}_2\sigma$, integration according to (4.1) yields per mole of a mixture

$$\frac{A_m^r}{RT} = \frac{A_m^{rh}}{RT} - \text{const.} \frac{\rho_m}{RT}\sum_i\sum_j x_i x_j \bar{u}_{ij}\bar{g}_{ij}\sigma_{ij}^3 - \ldots \tag{4.21a}$$

If all the peaks of $g(r)$ are equal, $\bar{g}_{ij} = \bar{g}_{ii} = \bar{g}_{jj}$, we obtain

$$\frac{A_m^r}{RT} = \frac{A_m^{rh}}{RT} - \text{const.} \frac{\rho_m}{RT}\sum_i\sum_j x_i x_j \bar{u}_{ij}\sigma_{ij}^3 - \ldots \tag{4.21b}$$

which is identical with the VDW approximation. In fact, the "constant" in (4.20) is a complicated function of density and temperature. According to Eq. (4.21), the SW approximation (4.15) is exactly valid only if the component molecules have identical size. However, when one or both components of a binary system are large molecules, the VDW is not better than the SW approximation because in both cases the distances r are calculated from the centers of the molecules. There exists some experimental evidence that for mixtures containing molecules of the size of propane or larger, the average interaction energy should be calculated from surface-to-surface distances and interactions. The theory of such interactions may start either from (4.15) or (4.21a). The success of the SW or the VDW approximations compared with that of the Mie's potential is surprising because it seemed that the latter reflected the true potential curve more closely. The evidence that the potentials approaching the SW conditions are better endorses the *cell model* of the liquid state in which the potential is uniform and a molecule is allowed to roam freely within a cell. The foundations of this model are due to Lennard-Jones and Devonshire (1937) and are lucidly outlined by Prigogine, Bellemans, and Mathot (1957), Hirschfelder, Curtiss, and Bird (1954), and Hill (1960). It was further developed by Kohler and Fischer (1967) and by Adams and Matheson (1972). This model may be valid for crystals and liquids near the triple point, but it is not necessary for the development of an equation of state.

    Each of the four approximations conforms to assumptions (4.3) and (4.5). Thus, the failure of the Lennard-Jones approximation does not mean that there are any deviations from the principle of corresponding states due to nonrandom mixing or other reasons. Moreover, we do not know a priori whether the form of $u(r)$ for, say, n-heptane differs significantly from that of argon. Further, the radial distribution function may not be exactly equal to $g^h(r)$ as in (4.1). It may be significantly affected by the shape of the molecules and by noncentral forces. Briefly, we are unable to pinpoint precisely the reasons for any deviations from the principle defined by (4.3) and (4.5).

    We shall now study the dependence of A^r on the reduced density and the reduced temperature to establish a more useful principle of corresponding states. Throughout this book the two variables are defined as

$$T^* = kT/\bar{u} \tag{4.22}$$

and

$$\rho^* = 1/V^* = V^o_m/V_m = \rho_m/\rho^o_m = N_o(\sigma^e)^3/(2^{1/2}V_m) \tag{4.23}$$

where V^o_m is the molar *random close-packed volume* and σ^e is the *effective* collision diameter of the molecules. It is a function of temperature (Chap. 6). We shall also use the variable

$$\xi = \frac{1}{6}\pi N_o(\sigma^e)^3/V_m = 0.74048\,\rho^* \tag{4.24}$$

which is the volume of N_o molecules divided by V_m. The superscript e will usually be deleted.

    The development of the various terms in $A^r(T^*, \rho^*)$ is reviewed below and in Chap. 6.

B. Systems of Small Molecules

At first, we assume that the molecules are small and nearly spherical, and consider the effect of multipoles on A. The theory is outlined by Rowlinson (1969), Stell, Rasaiah, and Narang (1972, 1974), and Rushbrooke, Stell, and Hoye (1973). In this theory it is assumed that the orientational part of $u(r)$ can be treated as a perturbation of the spherically symmetric potential. Accordingly,

$$A^r = A^L + A^{(2)} + A^{(3)} + \ldots \tag{4.25}$$

The term A^L, due to the London energy, contains $A^{(1)}$ which, as shown by Pople (1954), is equal to zero for spherically symmetric molecules; $A^{(2)}$ and $A^{(3)}$ are the perturbations due to pair and three-body interactions, respectively. Per particle,

$$A^{(2)}/N = \mu^4 f_2^\mu + \mu^2 Q^2 f_{1,1}^{\mu Q} + Q^4 f_2^Q + \text{octopole terms} \tag{4.26}$$

where

$$f_2^\mu = -\frac{\rho}{6kT} \int g_{oo}(r)\, r^{-6}\, dr; \tag{4.27}$$

$$f_{1,1}^{\mu Q} = -\frac{\rho}{2kT} \int g_{oo}(r)\, r^{-8}\, dr; \tag{4.28}$$

$$f_2^Q = -\frac{7\rho}{10kT} \int g_{oo}(r)\, r^{-10}\, dr; \tag{4.29}$$

and $g_{oo}(r)$ is the pair distribution function of the reference (unperturbed) system.

Again, $A^{(3)}$ contains the functions f_3^μ, $f_{1,2}^{\mu Q}$, $f_{2,1}^{\mu Q}$, etc.; the first one, according to Stell et al. (1974), is

$$\mu^6 f_3^\mu = \frac{\mu^6}{54}\left(\frac{\rho}{kT}\right)^2 \int g_{ooo}(r)\, u(r_{12}, r_{13}, r_{23})\, d\mathbf{r_2}\, d\mathbf{r_3} \tag{4.30}$$

where $g_{ooo}(r)$ is the three-particle distribution function of the reference fluid, and $u(r)$ is given by the Axilrod-Teller (1943) equation

$$u(r_{12}, r_{13}, r_{23}) = \frac{1 + 3\cos\alpha_1 \cos\alpha_2 \cos\alpha_3}{(r_{12}r_{23}r_{13})^3} \tag{4.31}$$

in which α_i are the interior angles of the triangle 1-2-3. Since (4.25) appears to be a very slowly converging series, Narang (1972; see Rushbrooke et al., 1973) proposed to express it by a Padé approximant:

$$A^r = A^L + A^{(2)}/(1 - A^{(3)}/A^{(2)}) \tag{4.32}$$

The approximant amounts to summing the series as if it were a geometrical progression, and it appears to be an excellent approximation.[†] It is interesting that while (4.25) is good for moderately polar fluids but fails for H_2O, NH_3, and acetone, the Padé approximant holds even for the fluids with the strongest dipole or quadrupole moments observed in nature. This was demonstrated by Gubbins and Twu (1977) by a comparison of the internal energy obtained by a molecular dynamics computer simulation of molecules with London energy plus electrostatic potentials of varying strength. These "experimental" data were determined by Wang and coworkers (1974) and by Haile (1976). The influence of dipoles on U/RT and P of dense gases was also determined by Yao, Greenkorn, and Chao (1982) by means of Monte-Carlo simulation.

Twu, Gubbins, and Gray (1975, 1976) and Flytzani-Stephanopoulos, Gubbins, and Gray (1975) extended this theory to mixtures. The expressions for $A^{(2)}$ and $A^{(3)}$ involve the state variables (T, ρ, x_i), the intermolecular potential parameters of the references $\bar{u}_{oo}^L$ and σ_{oo}, and certain relations for the integrals in (4.27) to (4.30) denoted by J or K. They are as follows:

For a dipolar fluid:

$$A^{(2)} = -\frac{2\pi N\rho}{3kT}\sum_i\sum_j x_i x_j \left(\frac{1}{\sigma_{ij}^3}\right)\mu_i^2\mu_j^2 J_{ij}^{(6)} \tag{4.33}$$

and

$$A^{(3)} = +\frac{32\pi^3 N}{135}\left(\frac{14\pi}{5}\right)^{1/2}\left(\frac{\rho}{kT}\right)^2$$

$$\times \sum_i\sum_j\sum_k x_i x_j x_k \left(\frac{1}{\sigma_{ij}\sigma_{jk}\sigma_{ik}}\right)\mu_i^2\mu_j^2\mu_k^2 K_{ijk}\binom{333}{222} \tag{4.34}$$

[†]For more details about Padé approximants, see Baker (1965) and Allen and coworkers (1975).

For a quadrupolar fluid:

$$A^{(2)} = -\frac{14\pi N\rho}{5kT}\sum_i\sum_j x_i x_j\left(\frac{1}{\sigma_{ij}^7}\right)Q_i^2 Q_j^2\, J_{ij}^{(10)} \tag{4.35}$$

and

$$A^{(3)} = +\frac{144\pi N\rho}{245(kT)^2}\sum_i\sum_j x_i x_j\left(\frac{1}{\sigma_{ij}^{12}}\right)Q_i^3 Q_j^3\, J_{ij}^{(15)}$$

$$+\frac{32\pi^3 N}{2025}(2002\pi)^{1/2}\left(\frac{\rho}{kT}\right)^2\sum_i\sum_j\sum_k x_i x_j x_k$$

$$\times\left(\frac{1}{\sigma_{ij}^3\sigma_{jk}^3\sigma_{ik}^3}\right)Q_i^2 Q_j^2 Q_k^2 K_{ijk}\binom{555}{444} \tag{4.36}$$

The dimensionless integrals $J_{ij}^{(n)}$ and $K_{ijk}(\ldots)$, obtained by molecular dynamics and expressing the distribution functions, are universal functions of T^* and ρ^*:

$$\ln\left|J_{ij}^{(n)}\text{ or }K_{ijk}\right| = A^{(n)}(\rho^{**})^2\ln T^{**} + B^{(n)}(\rho^{**})^2$$

$$+ C^{(n)}\rho^{**}\ln T^{**} + D^{(n)}\rho^{**}$$

$$+ E^{(n)}\ln T^{**} + F^{(n)} \tag{4.37}$$

where the reduced variables are $\rho^{**} = \rho\sigma_{oo}^3 = 1.41421\rho^*$ and $T^{**} = kT/\bar{u}_{oo}^L$.

In the initial calculations for model systems, the authors have chosen as the reference substance argon + krypton mixtures of the same composition as the polar fluid. The values obtained by a Monte-Carlo study for the Lennard-Jones potential, quoted by the authors, are ($i = Ar$, $j = Kr$):

pair	$\dfrac{\bar{u}^L/k}{K}$	$\dfrac{\sigma}{\text{Å}}$	
ii	119.8	3.405	
ij	139.9	3.519	
jj	167.0	3.633	(4.38)

The values of σ_{oo} and $\bar{u}^L_{oo}$ in (4.37) were calculated by using the van der Waals approximation. The values of the constants $A^{(n)}$, $B^{(n)}$, etc. in (4.37) are different for each of the integrals $J^{(6)}$, $J^{(10)}$, etc. in (4.33) to (4.36), and have been tabulated by Twu and coworkers (1976). In order to apply Eq. (4.32), we must know accurate values for A^L. In the initial model calculations, Twu and Gubbins used a multiparameter empirical equation for argon.

With $A(T^*, \rho^*, x_i, \ldots)$ given by (4.32), the liquid-vapor equilibrium pressures, volumes, and enthalpies are calculated by using the relations given in Chaps. 1 and 2. The reference values in (4.38) are useful only in model calculations. For real systems, they must be obtained by a tedious iteration procedure in which, given μ and Q of the components, $\bar{u}^L_{oo}$ and σ_{oo} are varied until the vapor pressures and liquid densities of the pure components agree with the observed values, using the Gibbs equilibrium conditions. In this manner, Gubbins and Twu (1977, 1978), Twu and Gubbins (1978), and Calado and coworkers (1978) calculated the liquid-vapor equilibria for the Xe + HCl and the Xe + HBr systems in excellent agreement with the experimental data determined by Calado et al. (1975). Some of the results are shown in Figs. 4.1 and 4.2. In these calculations, the anisotropic *overlap potential*, due to overlapping charges of the molecules (repulsion), and the electrostatic induction terms were taken into account. These terms may be absorbed into A^L in any approximate considerations. The generalized relations are given by Gray et al. (1978) and Moser et al. (1981).

It is important to note that the temperatures indicated in Fig. 4.1 are not far from the upper critical solution temperature $T^{ucs} = 155$ K of the Xe + HCl system, below which two liquid phases are formed. Hildebrand and Scott (1950) believed that ordering effects in solutions may exist at most a few degrees above T^{ucs}, where the phenomenon of critical opalescence appears. The foregoing results seem to confirm this point of view. Eqs. (4.33) to (4.36) are for a random mixture and yield correct results simultaneously at the three temperatures. However, the shift of the isotherms is not very sensitive to the value of H^E, which is most strongly affected by the ordering effects. A comparison of the values of H^E is necessary to judge the presence or the absence of order in mixtures.

The deviations from the principle of corresponding states are here

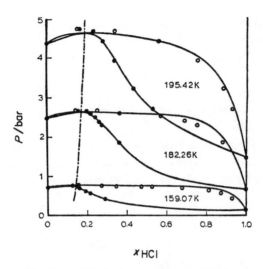

FIG. 4.1.   Vapor-liquid equilibria for the Xe + HCl system from theory (lines) and experiment (points). The dash-dot line is the azeotropic locus. Reproduced with permission from *J. Chem. Soc. Faraday Trans. I* 74:893.

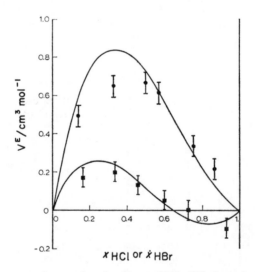

FIG. 4.2.   Excess volumes for the Xe + HCl (filled circles) and the Xe + HBr system (filled squares) compared with theory (lines). Reproduced with permission from *J. Chem. Soc. Faraday Trans. I* 74:893.

partly due to the three-body interactions, which cannot be neglected in accurate calculations. Even if they are neglected, $A^{(3)}$ vanishes for $\mu\text{-}\mu$ interactions but not for $Q\text{-}Q$ interactions. If Q_i and Q_j have opposite signs, $A^{(3)}$ in (4.36) is negative, and the total A is more negative than in the case of identical signs for Q_i and Q_j or in the absence of quadrupole moments. This is a source of negative deviations from an ideal system and, in extreme cases, of negative azeotropy. The hypothesis that negative deviations are due to opposite signs of quadrupole moments was first advanced by Rowlinson (1969), and (4.36) confirms it. The great variety of phase diagrams depending on the magnitude of $A^{(2)}$ and $A^{(3)}$ relative to A^L, and on the sign of $A^{(3)}$, is demonstrated by Twu and his colleagues (1976).

C. Systems of Large Molecules

Much less is known about A^L. According to the electrostatic theory, the dipole and the quadrupole contributions to A rapidly decrease with increasing size of the molecules. We shall now consider systems of large, nonspherical molecules in which the multipole contributions are either negligibly small or zero, so that $A^r = A^L$.

For argon, we need three parameters: $\bar{u}^o$, σ, and the temperature dependence of σ determined by a function with one constant—say, C. According to the theory of noncentral forces, a fourth parameter, determining the temperature dependence of $u(r)$—say, ω or η/k—is necessary to evaluate $A^L(T^*, \rho^*)$ of nonspherical molecules. That is, the equation of state and $A^L(T^*, \rho^*)$ for argon will also hold for these molecules, but $\bar{u}/k$ in T^* will be a function of T.

These simple assumptions may eventually fail for large molecules, particularly chain molecules, and this is one of the main unsolved problems of the theory of fluids.

The earlier attempts are extensively reviewed by Leland and Chappelear (1968). McGlashan and Potter (1962) showed that the second virial coefficients of n-alkanes, which they measured for the homologues up to $n\text{-}C_8H_{18}$, are well correlated by the empirical equation

$$\frac{\beta}{V^c} = \left(\frac{\beta}{V^c}\right)_o - 0.0375(m_c - 1)(T^c/T)^{4.5} \tag{4.39}$$

where $(\beta/V^c)_o$ are the values for small spherical molecules at the same T^c/T, and m_c is the number of carbon atoms. This equation does not mean that n-alkanes exhibit deviations from the principle of corresponding states. Since the acentric factor ω or η/k of n-alkanes is a continuous function of m_c, Eq. (4.39) expresses the effect of noncentral forces in n-alkanes. A more general relation, not restricted to n-alkanes, would result upon replacing m_c by an equivalent function of ω.

Prigogine and coworkers (1957) have extended the principle to chain molecules by the use of a parameter c_r, which depends on the ratio of the number of chain *links* (r) to the number of external degrees of freedom per molecule. These degrees of freedom arise from the motion of the molecules and from independent motions of the chain links. Bhattacharyya, Patterson, and Somcynsky (1964) applied this concept to the evaluation of the excess functions of mixtures.

More recently, Flory (1965, 1970), Abe and Flory (1964, 1965), Orwoll and Flory (1967), and Eichinger and Flory (1968) developed an equation of state for chain molecules, the first equation properly *accounting for the repulsion at high densities*,[†] and applied it to mixtures of such molecules with very good results. However, it is limited to the liquid state far from the critical point. It does not reduce to the perfect gas laws when $\rho \to 0$, nor does it permit evaluation of $\beta(T)$.

Leach, Chappelear, and Leland (1966) and Leach (1967) expressed the contribution of noncentral forces by means of the *shape factors* $\theta_{ij,oo}$ and $\phi_{ij,oo}$. These parameters bear double subscripts, one for the given fluid and another for the reference. They can be defined in two ways: theoretical and empirical. The empirical definition, which proved to be more useful, is (for a pure fluid):

$$f_{ii,oo}(V/V_i^c, T/T_i^c) = (T_i^c/T_o^c)\,\theta_{ij,oo}(V/V_i^c, T/T_i^c); \qquad (4.40)$$

$$h_{ii,oo}(V/V_i^c, T/T_i^c) = (V_i^c/V_o^c)\,\phi_{ij,oo}(V/V_i^c, T/T_i^c) \qquad (4.41)$$

For small spherical molecules $\theta_{ij,oo} = 1$, $\phi_{ij,oo} = 1$, f_{ii} and h_{ii} become constants for a given species, and the two relations become identical with (4.2). The relations for the shape factors of hydrocarbons relative

---

[†]The equation of state for hard spheres, developed at about the same time, was not supposed to hold for chain molecules.

to methane as the reference fluid o, obtained from the data for Z and G^r/RT in the range $0.6 < T/T^c < 1.5$, are

$$\theta_{ii,oo} = 1 + (\omega_i - \omega_o) \left[0.0892 - 0.8493 \ln(T/T_i^c) \right.$$

$$\left. + \left(0.3063 - \frac{0.4506 \, T_i^c}{T} \right) \left(\frac{V}{V_i^c} - 0.5 \right) \right]; \qquad (4.42)$$

$$\phi_{ii,oo} = \frac{Z_o^c}{Z_i^c} \left\{ 1 + (\omega_i - \omega_o) \left[0.3903 \left(\frac{V}{V_i^c} - 1.0177 \right) \right. \right.$$

$$\left. \left. - 0.9462 \left(\frac{V}{V_i^c} - 0.7663 \right) \ln(T/T_i^c) \right] \right\} \qquad (4.43)$$

where Z^c are the compressibility factors at the critical point. When $V/V_i^c \geq 2$ or $V/V_i^c \leq 0.5$, the value of V/V_i^c in both equations is to be set equal to the constant value, 2 or 0.5, respectively. For methane, the authors used $\omega_o = 0.005$. According to Rowlinson (1974), if the reference fluid is one for which ω is approximately zero, the method is restricted to substances for which ω is less than about 0.25. This means that errors in the calculated thermodynamic properties may appear for n-hexane ($\omega = 0.290$) or longer chains. The *PVT* and the thermal properties of mixtures are calculated by starting with (4.6) and using the basic relations of Chap. 1. These relations have been rederived by Rowlinson and Watson (1969) in forms suitable for the application of the shape factors. Watson and Rowlinson (1969), Gunning and Rowlinson (1973), Teja and Rowlinson (1973), and Mollerup (1974, 1975) have shown that the method yields good results for liquid-vapor equilibria at high pressures, gas-liquid critical constants, and liquid densities of binary and ternary systems. Also, limited miscibility of two fluid phases below or above the gas-liquid T^c was predicted for some systems in qualitative agreement with observed data. Since the properties of mixtures are less sensitive to errors in an equation of state than are those of pure fluids, the method is also suitable for molecules somewhat larger than hexane.

Recently, Glowka (1977) compared graphs of G^r/RT of various substances as functions of $\log(P/P^c)$ and $\log(T/T^c)$ and found that they can be nearly exactly superimposed *by rotating* them around the critical point. The resulting relations are

$$f_{ii,oo} = \frac{T_i^c}{T_o^c} \left(\frac{P}{P_o^c}\right)^{(-\sin \gamma_{io})} \left(\frac{T}{T_o^c}\right)^{(\cos \gamma_{io} - 1)}$$

(4.44)

and

$$h_{ii,oo} = \frac{T_i^c P_o^c}{T_o^c P_i^c} \left(\frac{P}{P_o^c}\right)^{(-\sin \gamma_{io} - \cos \gamma_{io} + 1)} \left(\frac{T}{T_o^c}\right)^{(-\sin \gamma_{io} + \cos \gamma_{io} - 1)}$$

(4.45)

where γ_{io} is the angle of rotation with respect to the contour map for argon as the reference substance ($\sin \gamma_o = 0$). The Gibbs energies are reproduced in this way more accurately than by means of the shape factors. Massih and Mansoori (1983) developed a theory which indicates that the molecular shape factors depend only on temperature; they believe that the density dependence in (4.42) and (4.43) may be due to the use of an empirical equation of state.

Chen and Kreglewski (1977) tested the principle of corresponding states by means of an accurate equation for A_m^r in which the repulsion term A^h of (4.1) is a theoretical one for convex bodies of any shape, and the integrals involving the distribution functions are replaced by a power series of $1/T^*$ and ρ^*, $A^L(T^*, \rho^*)$. In accordance with the theory of noncentral forces, η^L/k was proportional to ωT^c. Since ω is determined by the vapor pressure of a liquid, η^L/k was that of a high-density fluid at $T/T^c = 0.7$. Also the remaining constants $\bar{u}^o/k$ and σ were obtained from the PVT data and internal energy data of a fluid at high densities. The second virial coefficients ($\rho = 0$) predicted with these constants for hydrocarbons up to pentane are in excellent agreement with the experimental data. Thus all the fluids up to C_5H_{12} behave like argon, the only difference being that $\bar{u}^L/k$ is a function of T, $\bar{u}^L = \bar{u}^o(1 + \eta^L/kT)$.

Recently, the author determined the constants σ, $\bar{u}^o/k$, and η^L/k from high-density data for several fluids with large molecules and used them for extrapolations to $\rho = 0$. The errors in the predicted values of $\beta(T)$ of benzene and n-hexane are small, but those for n-heptane, shown in Fig. 4.3, are considerable. Curve 1 is obtained by using the constants of a high-density fluid ($V^o = 93.54$ cm^{3}mol^{-1}, $\bar{u}^o/k = 449.72$ K, $\eta/k = 130$ K). Curve 2 was calculated by retaining the same values of V^o and $\bar{u}^o/k$ but with a new value, $\eta/k = 185$ K, to fit approximately the experimental points.

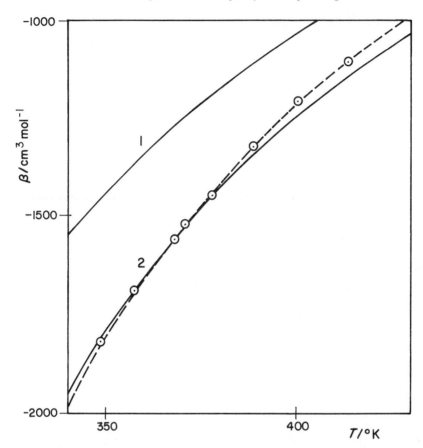

FIG. 4.3.   The second virial coefficients of *n*-heptane at low temperatures T^*, determined by McGlashan and Potter (1962). Curves 1 and 2 were calculated from an equation of state as explained in the text.

The slope of curve 2 is incorrect. The observed curve tends to higher values at high temperatures—where $\beta(T)$ tends to $2/3\,\pi N_o \sigma^3$—indicating that *the effective collision diameters of chain molecules in a dilute gas are larger than those calculated from the properties of the liquid state*. In other words, and quite obviously, the radius of gyration is larger in the gas than in the liquid phase.

The effect of the density of the fluid on $\bar{u}/k$ is much more significant. Almost certainly, η/k of chain molecules longer than pentane *is larger in the gas than in the liquid phase* because of the restricted rotation of the molecules. The rate of collisions in a dilute gas is proportional to the square of density. This argument leads to the approximate relation

$$\eta = \eta^o[1 - \kappa(\rho^*)^2] \tag{4.46}$$

where η^o is the value at $\rho^* = 0$, from $\beta(T)$ data, and κ is a characteristic constant for a given species and may be called the *hindering factor* (or coefficient). For small molecules up to pentane, $\kappa \approx 0$. For larger molecules, κ is expected to increase up to such a value that $\kappa\rho^* = 1$ at the triple-point density of the liquid. This value of density, estimated from accurate equations of state, is close to $\rho^* = 0.7$, so the largest anticipated value of κ is $2.04 \approx 2$. The value of η/k is usually estimated from the acentric factor ω. The acentric factor is determined at $T/T^c = 0.7$. At this T/T^c, the liquid reduced density of argon is $\rho^* = 0.519$. The value appears to be different for liquids with large molecules, but we know only an uncertain average $\langle\sigma\rangle$. By taking $\rho^* = 0.519$ for all liquids at which ω and η/k are determined, we obtain for *n*-heptane $\eta/k = 130$ K at $\rho^* = 0.519$ and $\eta/k \approx 185$ K at $\rho^* = 0$ (Fig. 4.3), and from (4.46), $\kappa \approx 0.57$. For benzene, the values are 72 K and 88 K, respectively, and we obtain $\kappa \approx 0.35$.

In the next chapter, we shall consider an alternative, namely, that $\eta = \eta^o$ is constant and the temperature dependence of $\bar{u}/k$ for large molecules is moderated at high densities by a term $\Delta/(kT)^2$ (see Eq. 5.10a).

The assumption made in the theories of chain molecules, developed by Prigogine and Flory, and their coworkers, is that $u(r)$ is independent of T, which seems to contradict the theory of noncentral forces. In view of the foregoing results, however, the assumption that $\eta/k = 0$ is allowed at high densities and for long chains, for which we expect that $\kappa \to 2$. Koningsveld (1970) has shown that $\bar{u}_{ij}/k$ of polymer solutions depends on T, but it may be partly due to η/k of the solvent not being equal to zero. In this case also η_{ij}/k is non-zero (Eq. 5.7).

The discussion in this chapter may be summarized as follows: (1) Most systems deviate from the classical principle of corresponding states, but not necessarily because of the failure of (4.5). (2) The deviations are due mostly to the differences in the distribution functions of various fluids as functions of ρ^* and T^*. (3) *There exists the follow-*

ing function, $A^r(T^*, \rho^*)$, valid for all fluids (non-electrolytes), which replaces the classical principle of corresponding states:

$$A^r(T^*, \rho^*) = A^L(T^*, \rho^*) + A^{(2)}(T^*, \rho^*) + A^{(3)}(T^*, \rho^*) \quad (4.47)$$

where, including large molecules, $\bar{u}^L = \bar{u}^L(T, \rho^*)$ and $\sigma = \sigma(T^*, \rho^*)$. Chap. 6 is devoted to the development of $A^L(T^*, \rho^*)$. In mixtures of molecules of different sizes or interaction energies, certain ordering effects may appear that affect the chemical potentials (Chap. 5).

CHAPTER 5

The Combining Rules and the
Ordering Effects in Mixtures

The thermodynamic properties of mixtures are extremely sensitive to the values of the "mixed" interactions, $u_{ij}(r)$, relative to those for the pure components, $u_{ii}(r)$ and $u_{jj}(r)$. A theory of fluids that predicts A^r, Z, and U^r with errors not exceeding, say, ± 2 percent is very satisfactory. However, the same error in the predicted value of $\bar{u}_{ij}$ may greatly affect the phase diagram of a binary system. If the predicted value is 2 percent low, the phase diagram may, for example, exhibit an azeotropic point or even a separation into three phases, liquid-liquid-vapor, while the diagram of the real system shows only two, liquid and vapor phases, and no azeotrope.

The parameters $\bar{u}_{ij}^o$, σ_{ij}, ω_{ij}, η_{ij}, etc., that determine the Helmholtz energy of the mixture $A_m^r(T^*, \rho^*, x_i, x_j, \ldots)$ are calculated from those for the pure components by means of the *combining rules*. Given the great variety of molecular shapes and kinds of interactions encountered when dealing with large molecules, it is unlikely that universally valid *accurate* combining rules will ever be derived. They may be essentially correct and useful in the discussion of phase diagrams but not good enough to obtain the *predicted* values required in chemical engineering.

Even if a combining rule is correct, the results obtained may be completely wrong because of improper averaging of the interactions. In the four approximations treated in Chap. 4, the interactions are weighted by means of the mole fractions, but it is doubtful that this is adequate when the molecular sizes of the components are very different.

Further, the behavior of a real fluid may be affected by certain ordering effects not taken into account by the combining rules. The problem of weighting the interactions is discussed in Section B and the ordering effects in Section C of this chapter.

A. The Combining Rules for $\bar{u}_{ij}$ and σ_{ij}

We begin with the combining rules for the London energy $\bar{u}_{ij}^L$ acting between small spherical molecules. It is usually assumed that $h\nu$ in (3.1) is equal to the ionization potentials.[†] Since they are similar for various organic substances, they may eventually be eliminated, leading to the rules: $\bar{u}_{ii}^L \sim p_i^2$, $\bar{u}_{jj}^L \sim p_j^2$, and

$$\bar{u}_{ij}^L \approx (\bar{u}_{ii}^L \, \bar{u}_{jj}^L)^{1/2} \tag{5.1}$$

Certain corrections to (5.1), due to neglecting the inequality of the ionization potentials, were derived by Thomsen (1971). Other combining rules in which (5.1) is the principal part are reviewed by K. M. Smith and his coworkers (1977) and by Huff and Reed (1963). In view of the uncertainty of the meaning and the values of $h\nu$, it is better to eliminate these quantities as follows. The interaction between hard spheres may vary according to (3.13), with r^{-6} at all distances $r > \sigma$. This can be expressed by the two following relations:

$$u_{ij}(r) = \infty \quad \text{at} \quad r \le \sigma \tag{5.2}$$

and

$$u_{ij}^L(r) = -\bar{u}_{ij}^L(\sigma_{ij}/r)^6 \quad \text{at} \quad r > \sigma \tag{5.3}$$

The potential expressed by these relations is known as the *Sutherland potential*. By comparing (3.1) and (5.3), we obtain for the pure components

$$h\nu_i = \frac{4}{3} \, \bar{u}_{ii}^L \sigma_{ii}^6 / p_i^2 \quad \text{and} \quad h\nu_j = \frac{4}{3} \, \bar{u}_{jj}^L \sigma_{jj}^6 / p_j^2 .$$

Upon substitution into (3.1), the quantities $h\nu$ cancel so that

$$\bar{u}_{ij}^L = 2 \left[\frac{1}{\bar{u}_{ii}^L} \left(\frac{p_i}{p_j} \right) \left(\frac{\sigma_{ij}}{\sigma_{ii}} \right)^6 + \frac{1}{\bar{u}_{jj}^L} \left(\frac{p_j}{p_i} \right) \left(\frac{\sigma_{ij}}{\sigma_{jj}} \right)^6 \right]^{-1} \tag{5.4}$$

This combining rule was derived by Kohler (1957). The Slater-Kirkwood (1931) formula for the London energy contains N_e/p_i, in-

---

[†]The ionization potentials, the electron affinities, and the methods of their determination are given by Gurvich et al. (1974) and by McDowell (1969).

stead of $h\nu_i$, where N_e is the number of electrons in the outer shell. As shown by Pitzer (1959), N_e appears to be different from the number of electrons in the outer shell. By eliminating N_e/p_i, we obtain the same combining rule, (5.4).

It can be shown that this combining rule, like the Lennard-Jones approximation for $\bar{u}_m$, greatly overstates the effect of molecular sizes.

In the square-well approximation, we have $F(\sigma_{ij}/r) = $ const. instead of $(\sigma_{ij}/r)^6$ in (5.3), and then

$$\bar{u}_{ij}^L = 2\left[\frac{1}{\bar{u}_{ii}^L}\left(\frac{p_i}{p_j}\right) + \frac{1}{\bar{u}_{jj}^L}\left(\frac{p_j}{p_i}\right)\right]^{-1} \quad \text{(small molecules)} \quad (5.5)$$

This rule was derived by Kreglewski and Chen (1978). It will be shown later that the values of G^E, H^E, and $\beta_{ij}(T)$, predicted by (5.5) for systems of small molecules with London energies only, are in excellent agreement with the observed data. Fender and Halsey (1962) derived a simple but less accurate rule, in which $\bar{u}_{ij}$ is the harmonic mean of $\bar{u}_{ii}$ and $\bar{u}_{jj}$, as follows: $\bar{u}_{ij} = 2[(1/\bar{u}_{ii}) + (1/\bar{u}_{jj})]^{-1}$. The effect of the collision diameters is not completely eliminated in (5.5). The polarizability of molecules with London energies only is an approximately linear function of σ^3 or the close-packed volume V^{o}.[†]

The combining rule corresponding to the VDW approximation is uncertain. In the theory of van der Waals the parameter a is proportional to $\bar{u}\sigma^3$, where $\bar{u}$ is the minimum value of an unknown potential $u(r)$. As shown in Chap. 4, the VDW approximation is much better than the LJ approximation. The combining rules are applied here to a_{ij}, usually $a_{ij} = k_{ij}^a(a_{ii}a_{jj})^{1/2}$, where k_{ij}^a is a constant, because nothing better may be derived when $u(r)$ is not known. Snider and Herrington (1967) used the following rule:

$$a_{ij} = \left[a_{ii}a_{jj}/(\sigma_{ii}^3\sigma_{jj}^3)\right]^{1/2}\left[\frac{1}{2}(\sigma_{ii} + \sigma_{jj})\right]^3$$

which is identical with (5.1) if $a_{ij} \sim \bar{u}_{ij}\sigma_{ij}^3$ and $\sigma_{ij} = (\sigma_{ii} + \sigma_{jj})/2$.

[†]For the methods of calculation of bond and molecular polarizabilities and the values for numerous compounds, the reader is referred to the papers by Denbigh (1940), Beran and Kevan (1969), Barnes et al. (1971), Sanyal et al. (1973), and the references given there. Tables of polarizabilities are given in the NBS Circular (1953) and in the Landolt-Bornstein Tabellen, vol. 1 (1960).

As mentioned in Chap. 4, the attraction term in the van der Waals equation of state, a/RTV, may be obtained by assuming a square well for $u(r)$ and a square "top" for $g^h(r)$ (Eq. 4.20). Hence, Eq. (5.5) is also the proper combining rule for $\bar{u}_{ij}$ in the van der Waals constant a_{ij}.

However, if we insist on coupling $\bar{u}_{ij}$ with σ_{ij} and operating with a_{ij}, the effect on the calculated properties of mixtures is as if $u_{ij}(r)$ were expressed by a hypothetical potential $\bar{u}_{ij}(\sigma_{ij}/r)^3$. This leads to a combining rule in which all the σ^6 in (5.4) are replaced by σ^3. This will hereafter be called the VDW approximation rule.

The rules for the VDW and SW approximations are compared with the experimental data for mixtures of noble gases in Table 5.1. The experimental values of $\bar{u}_{ij}$ are those selected by K. M. Smith and coworkers (1977) from molecular beam scattering, second virial coefficient, viscosity, and diffusion data. It appears that the geometric mean rule, (5.1), yields values that are much too high, whereas the values resulting from rule (5.5) are within the uncertainty of the experimental values, which (comparing with the values estimated by C. H. Chen et al., 1973), amounts to ± 10 percent. The rule for the VDW approximation is also satisfactory.

The results of quantum-mechanical calculations of $\bar{u}_{ij}\sigma_{ij}^6$ for interactions between noble gases, reported by Henderson and Leonard (1971), are less conclusive because of the uncertainty of σ_{ij}. They find that, always,

$$\bar{u}_{ij}^L\sigma_{ij}^6 < (\bar{u}_{ii}\bar{u}_{jj}^L)^{1/2}(\sigma_{ii}\sigma_{jj})^3$$

TABLE 5.1.   The values of $\bar{u}_{ij}^L/k$ for mixed interactions between noble gases (in Kelvins).

	Experimental	Eq. (5.5)	VDW[(a)]	Eq. (5.1)	k_{ij}^u Eq. (5.6)
He + Ar	30.2	29.0	33.0	38.6	0.782
He + Kr	30.2	28.8	35.2	45.6	0.662
He + Xe	28.0	25.3	33.8	54.3	0.516

NOTE: The values of $\bar{u}^L/k$ used for the pure components are as follows: He, 10.5 K; Ar, 140 K; Kr, 196 K; and Xe, 265 K.
(a) Eq. (5.4) with all σ^6 replaced by σ^3; σ_{ij}^3 was calculated from Eq. (5.12).

Since, always, $\sigma_{ii}\sigma_{jj}/\sigma_{ij}^2 < 1$, $\bar{u}_{ij}^L$ must be smaller than the geometric mean of $\bar{u}_{ii}^L$ and $\bar{u}_{jj}^L$.

It is usual to correct for the departures from (5.1) by using the formula

$$\bar{u}_{ij} = k_{ij}^u(\bar{u}_{ii}\bar{u}_{jj})^{1/2} \tag{5.6}$$

The superscript L is dropped here because this rule is used also for polar and quadrupolar systems, and k_{ij}^u is adjusted to fit one or more properties of the given binary system. As shown in the last column of Table 5.1, k_{ij}^u invariably decreases with the increasing ratio of the London energies $\bar{u}_{jj}^L/\bar{u}_{ii}^L$.

Kreglewski and Chen (1978) have found that rule (5.5) is valid only for systems of small molecules for which $\bar{u}_{ij}^L \approx \bar{u}_{ij}^o$ and $\eta/k \approx 0$. For systems in which at least one component exhibits noncentral forces stronger than those of ethane ($\eta^L/k = 19$ K), the rule begins to fail. We may improve the rule by applying the theory of Salem—namely, Eq. (3.4)—for close distances between chains. However, the rule so obtained is too inaccurate to be useful in practical applications.

In addition, we need a combining rule either for the total $\bar{u}_{ij}^L$ or for η_{ij}^L. In a random mixture, the part of the London energy depending on mutual orientations of the molecules is probably not affected by the presence of the molecules of the other species. Hence, we assume that

$$\eta_{ij}^L = (\eta_{ii}^L + \eta_{jj}^L)/2 \tag{5.7}$$

According to Tancrede, Patterson, and Lam (1975) and Tancrede and coworkers (1977), the interaction parameter X_{12}, used in the theory of chain molecules, contains X_{12}(chem) depending on the chemical nature of the components and X_{12}(orient) depending on their orientations. They obtained, per unit surface s_1 of component 1,

$$\frac{X_{12}(\text{orient})}{s_1} = \frac{1}{2}\text{const.}[J_{11} + J_{22} - 2(J_{11}J_{22})^{1/2}]$$

Since $X_{12}/s_1 \sim w_{11} + w_{22} - 2w_{12}$, where w_{ij} are certain interaction energies, the orientational part w_{12}(orient) appears to be proportional to $(J_{11}J_{22})^{1/2}$. The empirical quantities J_{ij} are related to molecular optical anisotropy by the theory of Kielich (1971) and Bothorel (1968), and it is not apparent from these relations that they vary with $(kT)^{-1}$ as required by the theory of noncentral forces. The quantity J_{ij}, then, is not directly related to η_{ij}^L in (5.7).

In the theories of long-chain molecules and polymers, it is usual to assume that $\bar{u}_{ii}$ is independent of the temperature and that it is the sum of interactions ε_{ii} between m_c segments of equal size. These segments are not distinguished, and certain average values of ε_{ii} are ascribed to each of them so that $\bar{u}_{ii} = m_{ci}\varepsilon_{ii}$.[†] Since η/k of each of these quasi-spherical segments is obviously zero, $\bar{u}_{ii}$ could also be regarded as independent of T. However, *the segments are connected*, and ε_{ii} are parts of $\bar{u}_{ii}$; therefore, it is better to assume that ε_{ii} depends on T to the same extent as does $\bar{u}_{ii}$. In the limiting case of long chains at high densities, $\eta/k \rightarrow 0$, as stated in Chap. 4.

For segments of equal size and, implicitly, equal polarizabilities, (5.5) simplifies to

$$\varepsilon_{ij}^o = 2\,\gamma_{ij}[(1/\varepsilon_{ii}^o) + (1/\varepsilon_{jj}^o)]^{-1} \tag{5.8}$$

where γ_{ij} is a coefficient. This rule was for the first time applied to segment interactions by Kreglewski and Kay (1969). The contributions of ε_{ii}, ε_{jj}, and ε_{ij} to $\bar{u}_m/k$ were weighed by means of surface fractions, and for systems with London energies only, γ_{ij} was equal to unity. The agreement of the predicted critical temperatures $T^c(x)$ and pressures $P^c(x)$ with the observed values is excellent. From this study it appears that crudely estimated values of ε of the *pure* components suffice for the studies of mixtures. Elaborate treatments taking into account the specific geometry of different molecules do not improve the results, because the failures are usually due to other sources of error.

In any case, the deviations from an ideal mixture due to the inequality of London energies of the components are small. For example, G^E/RT of a binary system at $x = 0.5$ seldom exceeds 0.08 (about $200\ \text{J} \cdot \text{mol}^{-1}$ at 298 K). The greatest influence on the deviations from an ideal system are the dipole and the quadrupole interactions, probably including the effect usually ascribed to the hydrogen bridges (bonds). These effects on A_m^r are given by functions of T^* and ρ^* different from those for the London interactions. The properties of mixtures, *relative* to those of the pure components, are not sensitive to

---

[†] *The group contribution method* in which different values of ε are ascribed to each atom or group of atoms will be considered in Chap. 9.

dependence on ρ^*. It is known that even very poor equations of state may yield good values of the excess functions. However, these functions are extremely sensitive to the dependence of A^r on $\bar{u}$ and to the combining rule for $\bar{u}_{ij}$. Eq. (4.25) may be rewritten, including the effect of η^L/k, as $A^r = A^o[1 + (A^{(\eta)} + A^{(2)} + A^{(3)})/A^o]$. We now assume that all the terms are identical functions of ρ^* and T^*, differing only in the parameters η^L/k, μ, Q, and so on, that determine $u(r)$. They are combined as follows:

$$\bar{u}_{ij} = \bar{u}_{ij}^o \left[1 + \frac{\eta_{ij}^L}{kT} + \frac{(\eta_{ii}^*\eta_{jj}^*)^{1/2}}{kT} - \frac{\eta_{ii}^{**}\eta_{jj}^{**}}{(kT)^2} \right];$$

$$(i = j \text{ or } i \neq j) \tag{5.9}$$

The product $\eta_{ii}^*\eta_{jj}^*$ is proportional to $A^{(2)}$, appears in the presence of dipole or quadrupole interactions or both, and is *always positive*. The product $\eta_{ii}^{**}\eta_{jj}^{**}$ is related to $A^{(3)}$ and, if the three-body interactions in (4.36) are ignored, is zero in the case of dipole moments but nonzero in the presence of quadrupoles Q. The signs of η_{ii}^{**} and η_{jj}^{**} are identical with those of Q_i and Q_j, respectively.

If one of the components is an *inert solvent for which* $\eta_{ii}^* = 0$ *and* $\eta_{ii}^{**} = 0$, then $\bar{u}_{ij} = \bar{u}_{ij}^L = \bar{u}_{ij}^o(1 + \eta_{ij}^L/kT)$. When the second component j is a dipolar or a quadrupolar fluid, large positive deviations from an ideal mixture will occur if $\bar{u}_{ii}^L$ is not very different from $\bar{u}_{jj}^L$. If, however, $\bar{u}_{ii}^L \gg \bar{u}_{jj}^L$ and the dipoles in j are weak, the deviations may be quite small.

There are numerous examples of such systems. In one of them, Xe + HCl (shown in Fig. 4.1), $\bar{u}^L$ of both the components are weak compared to $\bar{u}^{\mu\mu}$ and $\bar{u}^{QQ}$ in HCl, and the deviations from ideal properties are large. In another one, CCl_4 + $CHCl_3$, the London energies are strong compared to $\bar{u}^{\mu\mu}$ in chloroform. The excess Gibbs energies of this system, determined by McGlashan, Prue, and Sainsbury (1954), are small and equal at $x = 0.5$ to 106.0 (25°C), 102.0 (40°C), and 91.0 J · mol^{-1} (55°C). Carbon tetrachloride is often believed to be an inert solvent, but it is not, and the system with chloroform is rather complex.

When both components are dipolar or quadrupolar, their quadrupole moments have the *same sign*, and their values of $\bar{u}^L$ are similar, then the deviations from an ideal mixture are positive and not large, because the total $\bar{u}_{ij}$ is close to the geometric mean of $\bar{u}_{ii}$ and $\bar{u}_{jj}$. Accidentally, it may even happen that the system is practically ideal. An example is the ethylene bromide + propylene bromide system, which—as shown by Zawidzki (1900)—conforms to Raoult's law at 85°C. It can be predicted, however, that the system will not be ideal at all temperatures.

When the quadrupole moments have *opposite signs*, then the last term in (5.9) increases the value of $\bar{u}_{ij}$, whereas $\bar{u}_{ii}$ and $\bar{u}_{jj}$ decrease. This causes negative deviations from the properties of a system that would otherwise be ideal or exhibit positive deviations. The effect on the macroscopic properties is that G^E, H^E, and S^E decrease, whereas $T^c(x)$ increase. As shown in Chap. 9, such systems are quite common, and their behavior may have important industrial applications.

Eq. (5.9) may be tested by comparisons with observed values of saturated liquid and vapor densities, vapor pressures and enthalpies of vaporization ΔH^v of pure fluids, using an accurate equation of state. In practice, the two contributions, η^L/k and η^*/k are inseparable. Therefore, we write:

$$\bar{u}_{ij} = \bar{u}_{ij}^o\left(1 + \frac{\eta_{ij}}{kT} + \frac{\Delta_{ij}}{(kT)^2}\right); \qquad (i = j \text{ or } i \neq j) \qquad (5.10a)$$

where $\Delta_{ij} < 0$. For pure dipolar or quadrupolar fluids, such as SO_2 or CO_2, η/k is larger than the value calculated from acentric factor and ΔH^v varies with temperature faster at high reduced temperatures but slower at low temperatures than for non-polar fluids. Similar behavior is exhibited by long-chain molecules for which Δ_{ij} accounts for a tendency to ordering at low temperatures and high densities (in this case, Δ_{ij} should tend to zero with $\rho \to 0$). Attempts are being made to establish Δ_{ij} for pure fluids such as water, methanol, and carbon dioxide. At the present time, only the approximate values of η/k are known (Table 6.4) and the following relations are used: for the pure components, $\bar{u} = \bar{u}^o(1 + \eta/kT)$, and for mixed interactions,

$$\bar{u}_{ij} = \bar{u}_{ij}^o\left[1 + \frac{\eta_{ij}}{kT} + \left(\frac{\delta_{ij}}{kT}\right)^2\right]; \qquad (i \neq j) \qquad (5.10b)$$

The constant δ_{ij}/k is a parameter relative to zero for the pure components and so it is a (positive) excess over the unknown value of Δ_{ij}. Its importance is demonstrated in Chap. 8.

The next most important parameter in the theory of phase equilibria is the average collision diameter σ_m of the system. Among the relations for σ_m, the Lennard-Jones approximation (4.10) is inaccurate, to say the least. In the limit when $n \to \infty$, it does not tend to the relation for hard spheres even if all $f_{ij} = 1$.

If σ_m is given by (4.12), the relation common for the VDW and SW(O) approximations, there are at least two alternatives for σ_{ij}: namely, the Lorentz relation (discussed by van der Waals and Kohnstamm, 1908):

$$\sigma_{ij} = (\sigma_{ii} + \sigma_{jj})/2 \qquad (5.11)$$

and that due to van der Waals:

$$\sigma_{ij}^3 = (\sigma_{ii}^3 + \sigma_{jj}^3)/2 \qquad (5.12)$$

If $\sigma_{ii} \neq \sigma_{jj}$, σ_{ij}^3 calculated from (5.11) is always smaller than the value obtained from (5.12). Comparisons with the Monte-Carlo results for hard spheres, made by Salsburg and Fickett (1962), show that (5.12) is very nearly correct.

The relations for the mixing of hard spheres, derived by Lebowitz (1964), are coupled with the equation of state and will be given in Chap. 6. The volumes occupied by hard spheres, obtained by using (5.12), are only slightly larger than those obtained from Lebowitz's exact relations.

C. H. Chen, Siska, and Lee (1973) obtained accurate values of σ_{ij} for systems of noble gases by combining the results of molecular beam scattering and second virial coefficients. For the pure gases these are He $\sigma/\text{Å} = 2.65$; Ne, 2.75; Ar, 3.34; Kr, 3.64; Xe, 3.81. The values of σ_{ij} are compared with those calculated from (5.11) and (5.12) in Table 5.2. These values were determined for a "soft" potential.

K. M. Smith and his coworkers (1977) obtained for noble gases the following values of the distances r^{min} at which $u(r)$ reaches a minimum value: He $r^{min}/\text{Å} = 2.97$; Ar, 3.76; Kr, 4.01; Xe, 4.38. The distance r^{min} is proportional to σ. The values for mixed interactions are given in Table 5.3.

It appears that even the values obtained from (5.12), which are larger than those either from (5.11) or from Lebowitz's relations, are too small. The relations for hard spheres are derived for $u(r) = 0$ at distances $r > \sigma$, whereas the collision diameters in real systems are not entirely independent of $\bar{u}/k$. The problem is too complicated to be

TABLE 5.2.   The collision diameters for mixed interactions between the atoms of noble gases, $\sigma_{ij}/\text{Å}$.

	Experimental	Eq. (5.11)	Eq. (5.12)	k_{ij} (in Eq. 5.13)
He + Ne	2.73	2.70	2.70	1.03
He + Ar	3.09	3.00	3.03	1.06
He + Kr	3.27	3.14	3.22	1.05
He + Xe	3.61	3.23	3.33	1.27

TABLE 5.3.   The values of $r_{ij}^{\min}/\text{Å}$ for mixed interactions between the atoms of noble gases.

	Experimental	Eq. (5.11)	Eq. (5.12)
He + Ar	3.46	3.36	3.41
He + Kr	3.67	3.49	3.57
He + Xe	3.95	3.67	3.81

considered here. Details can be found in recent works by Fulinski and Jedrzejek (1974) and by Chang, Hwu, and Leland (1979). However, it is clear that among the simple combining rules, (5.12) is the best one. Deviations from the rule are also noted in the systems of nonspherical molecules for which σ is a certain average value. Empirically, rule (5.12) is corrected as follows:

$$V_{ij}^o = k_{ij}(V_i^o + V_j^o)/2 \tag{5.13}$$

where V_i^o and V_j^o are the molar close-packed volumes, and where k_{ij} may be a function of ρ^* and x_i but is usually a constant for the given system. The constant k_{ij} may be roughly estimated from the geometry of molecules following the methods developed by Kihara (1976), but empirical methods are preferable.

In all the approximations for f_x and h_x of mixtures, considered in Chap. 4, the three-body interactions were neglected without any *clearly* negative effects on the calculated properties of *binary systems*. It remains to be shown whether the knowledge of $\bar{u}_{12}$, $\bar{u}_{23}$, and $\bar{u}_{13}$ obtained from the properties of three binary systems suffices to evaluate $\bar{u}_m$ of a ternary or a multicomponent system, or whether a separate combining rule for $\bar{u}_{123}$ (or $\bar{u}_{ijk}$) is required.

As mentioned before, the effect of three-body interactions on the absolute values of $A^L(T, \rho)$ is not negligible; however, the effect on $A^L(T^*, \rho^*)$, obtained relative to argon as the reference substance, is small; and in mixtures, relative to the pure components, it may be entirely negligible.

The possibility of deviations from pairwise additivity of the terms in systems with multipole interactions cannot be ignored. The most favored mutual orientations between Q_1 and Q_2 in the $1 + 2$ system are disturbed upon the introduction of Q_3 into the system. Since the product $Q_i^2 Q_j^2 Q_k^2$ in (4.36) is always positive, the effect of ignoring this term in the evaluation of $A^{(3)}$ is that

$$A^{(3)}(\text{real}) > A^{(3)}(\text{calc}) \quad \text{if} \quad \text{sign } Q_i = \text{sign } Q_j;$$

$$A^{(3)}(\text{real}) < A^{(3)}(\text{calc}) \quad \text{if} \quad \text{sign } Q_i \neq \text{sign } Q_j \qquad (5.14)$$

Hence, in the first case the total $A_m^r(\text{real})$ will be less negative than $A_m^r(\text{calc})$, and vice versa in the second case. However, according to the Gubbins-Twu theory, the three-body terms contain only the pure component properties, and there is no need for any triple interaction parameters.

Gubbins and Twu (1977) calculated the liquid-vapor equilibria of the $CO_2 + C_2H_4 + C_2H_6$ system at high pressures and obtained very good results by using only the binary interaction parameters, in spite of relatively large values of Q with opposite signs in CO_2 and C_2H_4.

Morris and his colleagues (1975) determined G^E and H^E of all the binary systems formed by acetone (1), chloroform (2), methanol (3), ethanol (4), and n-heptane (5) and of the ternary $(1) + (2) + (3)$ and $(2) + (4) + (5)$ systems; they found some deviations from the results obtained by applying the binary interaction parameters to the ternary systems. The calculations were based on an empirical relation for G^E and H^E instead of an accurate relation for $A_m^r(T^*, \rho^*)$. The $(1) + (2)$ system has negative values of G^E and H^E and forms a negative azeotrope. Therefore, according to the Gubbins-Twu theory, acetone and chloroform must have quadrupole moments with opposite signs. Without knowing at least approximate values of μ and Q of the components, it is not possible to apply this theory to check whether the properties of these systems may be calculated without triple-interaction parameters. The experimental data for the foregoing and other systems, measured with high precision in the laboratory of van Ness, remain the basic data for future tests of the theory.

When the specific interactions are weak or absent, the pair-interaction terms suffice to calculate the properties of a ternary system. This was shown by Clarke and Missen (1974), who determined G^E of the $CH_3CN + C_6H_6 + CH_3 \cdot C_6H_{11}$ system. Analogous results were obtained in the author's calculations of the critical temperatures of ternary hydrocarbon systems.

B. The Statistical Weights of Interactions

Errors in the calculated thermodynamic properties of systems of large or small + large molecules are not necessarily due to the failure of a combining rule. If the significant range of interactions is approximately limited to $\sigma < r < 3/2\ \sigma$, as in the square-well potential, a part of the large molecule may not "feel" the presence of the small molecule, as schematically shown in Fig. 5.1 for an isolated pair of molecules. The range of the large molecule is calculated here from the diameter of the cylinder (instead of the average σ) because only the parallel orientations contribute to $\bar{u}^L(r)$; see the discussion of Fig. 3.1. It is clear that a combining rule for $\bar{u}_{ij}$, valid for molecules of similar

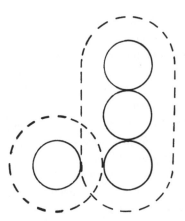

FIG. 5.1. A scheme of the interaction between a monomer and a trimer, each segment having the same width of square well.

sizes, may apparently fail for systems of molecules of similar chemical nature but differing in size.

According to the kinetic theory of gases (for example, Rowlinson, 1963), the total number of collisions between nonattracting hard spheres in unit of volume per unit of time is

$$z_{ij}^h = \frac{2}{1 + \delta_{ij}} \frac{N_i N_j}{N^2} \left(\frac{2\pi kT}{m_{ij}}\right)^{1/2} \sigma_{ij}^2 \rho_n^2 \tag{5.15}$$

where m_{ij} is the reduced mass

$$1/m_{ij} = 1/m_i + 1/m_j \tag{5.16}$$

and δ_{ij} is equal to unity if i and j are the same species and is zero otherwise. At a low density, constant T, and $u(r) = 0$, the frequency of collisions is proportional to the mole fractions, to σ_{ij}^2, and to $(1/m_{ij})^{1/2}$.

It is convenient to introduce certain g_i fractions defined as

$$\varphi_i = g_i^2 x_i \Big/ \sum_{h=1}^{m} g_h^2 x_h \tag{5.17}$$

According to (5.15), for a mixture of hard spheres at low densities,

$$g_i^2 = \sigma_{ii}/m_i^{1/4} \qquad [u(r) = 0, \text{ low } \rho_n] \tag{5.18}$$

or

$$g_1^2 = 1; \quad g_2^2 = (\sigma_{22}/\sigma_{11})(m_1/m_2)^{1/4}; \text{ etc.}$$

In the presence of intermolecular attraction, at very low densities $u(r)$ becomes the main factor affecting the number of collisions, since the range of $u(r)$ is greater than σ. The numbers of $u_{ii}(r)$ and $u_{jj}(r)$ centers are equal to N_i and N_j, respectively. Accordingly, the second virial coefficient of a mixture, as derived from statistical mechanics, is a quadratic function of the mole fractions x_i, regardless of the size or shape of the molecules. For a binary system, from (1.44),

$$\beta_m = x_1^2 \beta_{11} + x_2^2 \beta_{22} + 2x_1 x_2 \beta_{12} \tag{5.19}$$

This equation does not impose any restrictions on the various possible relations for $\bar{u}_{12}$ in β_{12}.

At high densities, more than two molecules are in steady contact, and $\bar{u}_m$ of a mixture is usually averaged by using the *contact fractions*, defined as

$$q_i = x_i s_i \left/ \sum_{h=1}^{m} x_h s_h = x_i \right/ \sum_{h=1}^{m} x_h (s_h / s_i) \qquad (5.20)$$

Deiters (1982) calculated the maximum number of hard spheres that can be attached to a central sphere of a different diameter; he found that in a binary system, s_2/s_1 is proportional to $(\sigma_2/\sigma_1)^{1.2}$. He also noted that the curves of G^E for systems of small spherical molecules become more symmetrical when plotted against contact fractions calculated for $(\sigma_2/\sigma_1)^{1.2}$.

The molecular interactions between globular, disk, or chain molecules are usually divided into the interactions between segments of the molecule. The idea was developed by Redlich and coworkers (1959), Bondi and Simkin (1960), McGlashan and coworkers (1961), and Patterson, Tewari, and Schreiber (1972). In these works, the thickness of the shell of interactions was not taken into account; that is, it was assumed to be zero. This shell makes the geometric features of a molecule less significant in interactions with other molecules. The equation of state for hard convex bodies contains a parameter $\alpha \geq 1$ (Chap. 6), which should increase with the deviations from spherical symmetry. The values of α obtained from the *PVT* properties of dense fluids are 1.0498 for neo-pentane and 1.0566 for *n*-pentane, the latter being much smaller than the geometrical value. Since the equation of state with an attraction term for real fluids is very accurate, the values of α obtained from it are probably realistic. The problem is less complicated when limited to chain molecules only, and we shall now consider the variation of $\bar{u}$ and V^o of *n*-alkanes with the number of carbon atoms m_c.

Flory, Orwoll, and Vrij (1964) have shown that V^o of *n*-alkanes is a linear function of m_c. The results are the same when V^o is fitted to the *PVT* properties of the liquid phase by the Beret-Prausnitz equation of state, considered in Chap. 6. The results of Beret and Prausnitz (1975) are shown in Fig. 5.2. The line would pass through the origin if V^o were plotted against $(m_c + 1)$. Flory assumed that the number of segments m_{ci} is proportional to V_i^o, and the interaction probabilities to the *site fractions* $s_i m_{ci} N_i$, where s_i is the part of the surface of a segment able to contact that of another molecule. If all $s_i = 1$, the site fractions simplify to the *volume fractions*.

The theory of Flory and his coworkers was successfully applied to mixtures of alkanes and alkanes + polyethylene. For example, when the differences in m_{ci} of the components are large enough, the systems

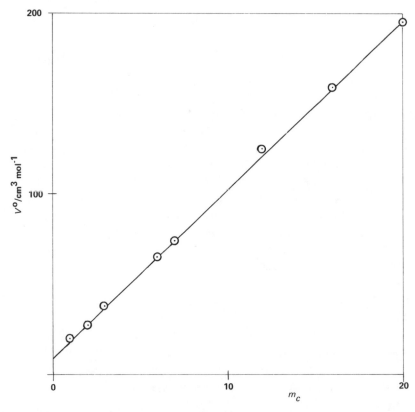

FIG. 5.2.   The values of V^o obtained by Beret and Prausnitz for *n*-alkanes up to C_{20} as a function of the number of C atoms.

may exhibit *lower* critical solution temperatures, T^{lcs}, above which the components are only partly miscible. The phenomenon is accurately predicted by this theory. This success is largely due to the equation of state (accurate at high densities), developed by Flory; it is not a proof that the site fractions properly measure the interaction probabilities. It is interesting that this theory accurately describes the dependence of H^E and V^E on x_i of systems of liquid metals, as shown by Brostow (1974).

By taking into account a certain flexibility of the chain, Kurata and Isida (1955) estimated that the effective length varies with $m_c^{2/3}$. Experimental results support the theory of Kurata and Isida. For example, Wall, Flynn, and Straus (1970) found that the enthalpies of vaporization ΔH^v of *n*-alkanes closely conform to the relation

$$\Delta H^v = 13.43 \, m_c^{2/3} - 0.08075 \, T + 12.22 \tag{5.21}$$

The proportionality between ΔH^v and $m_c^{2/3}$ was first predicted by Huggins (1939). For liquids at low reduced temperatures, $Z \rightarrow 0$; for the saturated vapor, $Z \rightarrow 1$; and $\Delta H^v/RT$ is practically equal to $(1 - U^r/RT)$ of the liquid. As shown in the next chapter, U^r/RT at *constant* ρ^* is nearly exclusively a function of $\bar{u}/kT$, and only these values may be used to estimate the dependence of $\bar{u}/k$ on m_c.

The segment interaction energies, estimated from the values of $\bar{u}/k$ obtained by Beret and Prausnitz (1975) for *n*-alkanes (C_1-C_{20}), are shown in Fig. 5.3. The volumes $V^{\ominus}$ at the temperature $T/T^c = 0.6$,

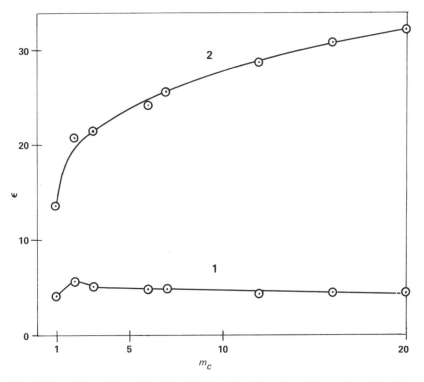

Fig. 5.3. The values of ε of *n*-alkanes plotted against the number of carbon atoms m_c. Curve 1: $\bar{u}/V^{\ominus}$; curve 2: $\bar{u}/(V^{\ominus})^{2/3}$.

given in the Appendix, were used here instead of V^o. In an analogy to the Scatchard-Hildebrand theory of solubility parameters, the proper values of ε are often supposed to be proportional to $\bar{u}/V^\bullet$. These values appear to be nearly constant for *n*-alkanes. According to Davies and Coulson (1952), the energies of interaction between the segments of two chains are enhanced by neighboring segments in their own chain. This is an argument in favor of the second alternative, $\varepsilon \sim \bar{u}/(V^\bullet)^{2/3}$, which should also be preferred in view of the foregoing discussion of the theory of Kurata and Isida. Moreover, the values of G^E/RT of the $C_1 + C_2$ and $C_1 + C_3$ systems are much larger than for the $C_6 + C_{16}$ system, which points to greater inequality of ε of the components in the first two systems than of those in the third system. The van der Waals approximation (4.11) fails for the above systems because it follows from theories in which $g(r)$ is derived for one-center molecules, whereas in molecules larger than methane two or more centers should be distinguished. This is a difficult problem in the case of globular or disk-shaped molecules such as neopentane or benzene. It is both easier and in most cases more proper to consider interactions between surfaces without paying too much attention to the molecular shapes.

All the calculations in this book will be based on surface interactions relative to the molecular surface of the smallest molecule in the system, $\varepsilon_{11} = \bar{u}_{11}/g_1^2$; $\varepsilon_{22} = \bar{u}_{22}/g_2^2$; etc., where $g_1 = 1$; $g_2 = (V_2^{oo}/V_1^{oo})^{1/3}$; etc., and where V_i^{oo} are the close-packed volumes at $T/K = 0$, weighted by the *surface fractions* defined as

$$\varphi_i = x_i g_i^2 \bigg/ \sum_{h=1}^{m} x_h g_h^2 \tag{5.22}$$

$$\varepsilon_m = \sum_i \sum_j \varphi_i \varphi_j \varepsilon_{ij}$$

The total interaction energy is given by

$$\bar{u}_m = \left(\sum_{h=1}^{m} x_h g_h^2 \right) \sum_i \sum_j \varphi_i \varphi_j \varepsilon_{ij}$$

$$= \sum_{h=1}^{m} x_h \bar{u}_{hh} + \sum_{i<j} \varphi_i x_j g_j^2 (2\varepsilon_{ij} - \varepsilon_{ii} - \varepsilon_{jj}) \tag{5.23}$$

For $g_1 = g_2 = \ldots = 1$, Eq. (5.23) becomes identical with (4.15). For a binary system,

$$\bar{u}_m = x_1 \bar{u}_{11} + x_2 \bar{u}_{22} + \varphi_1 x_2 g_2^2 (2\varepsilon_{12} - \varepsilon_{11} - \varepsilon_{22}) \tag{5.24}$$

For mixtures of inert components, such as $CH_4 + C_3H_8$ or $C_2H_6 + C_4H_{10}$, $\bar{u}_{12}$ obtained using the SW approximation (4.15) shows large deviations from the combining rule (5.5), whereas ε_{12} obtained using Eq. (5.24) approximately conforms to rule (5.8) with $\gamma_{ij} = 1$.

Delmas and Turrell (1974) and Huggins (1970, 1971) introduced differently defined surface fractions to correlate the excess enthalpies of systems of globular + chain molecules such as $Sn(C_mH_{2m+1})_4 + n$-alkane (C_5 to C_{16}). However, a surface parameter, analogous to s_i in the site fractions, has to be fitted here to the experimental data. If we also take into account disk-shaped molecules and our desire to obtain a relation for g_i without an empirical parameter, it is clear that only an approximate solution is possible.

The use of surface or site fractions is a temporary solution to the problems encountered in systems of large molecules. Besides molecular dynamics and Monte-Carlo studies of such systems, the first attempts were made to obtain the proper relation for the Helmholtz energy. From a practical point of view, probably the most promising approach was chosen by Boublik (1979 and earlier papers quoted here). If the difference between a soft and a hard convex body of any shape is neglected, the result is

$$\frac{A^r}{NkT} = \frac{A^{rh}}{NkT} + \frac{\rho_n}{2kT} \sum x_i x_j \int_0^\infty u_{ij}^p(l) \, g_{ij}^h(l) \, s_{ilj} \, dl \tag{5.25}$$

where A^{rh} is the residual Helmholtz energy of a system of hard convex bodies (see the next chapter); $u_{ij}^p(l)$ is the perturbation potential that for a given system depends only on the surface-to-surface distance l; $g_{ij}^h(l)$ is the distribution function of hard convex bodies; and s_{ilj} is the mean surface area of a body formed by the origin of core j when it moves around core i at a *constant* distance l. This surface area is given by

$$s_{ilj} = s_i + s_j + 8\pi l(I_{ci} + I_{cj}) + 4\pi l^2$$

where I_{ci} and I_{cj} are the $(1/4\pi)$ multiples of the mean curvature integrals, and s_i and s_j are the surface areas of the components. A method for evaluation of $g_{ij}^h(l)$ was worked out earlier by Boublik. The theory was applied by Sevcik, Boublik, and Biros (1978) to the cyclopentane + chloroform system, which was an unfortunate choice in view of the polarity of chloroform. The theory should be tested against the data for the methane (or nitrogen) + ethane, the propane, and the ethane + butane systems. Pavlicek and Boublik (1981) derived the second-order perturbation term in (5.25) and calculated the liquid-vapor equilibrium isotherm of the system $N_2 + C_2H_6$ at 200 K. The critical locus $P^c(x)$ curve of a system with a large ratio $\bar{u}_{22}/\bar{u}_{11}$ of the components usually exhibits a large maximum—in this case $P^c \approx 135$ bar at 200 K. The calculated P^c is about 60 bar higher than the observed value. The problem belongs to Chap. 8, after an introduction to critical phenomena. It is mentioned here because the errors are mostly due to *apparent* deviations from the rule $\sigma_{ij} = (\sigma_{ii} + \sigma_{jj})/2$, (Eq. 5.11), always used in the tests of theories of mixtures, or to the deviations from additivity of the close-packed volumes ($k_{ij} \neq 1$, in Eq. 5.13). Using an accurate equation of state and $k_{ij} = 1.10$ for the $N_2 + C_2H_6$ system, one obtains correct dew-point densities and compositions and a P^c value at 200 K only about 10 bar higher than the observed value. The deviations from the combining rules are caused here by factors other than differences between σ_{ii} and σ_{jj} because similar errors in the calculated critical pressures are noted for the $CH_4 + H_2S$ system (Chap. 7). It is an empirical fact that *the close-packed volumes of mixtures of fluids with very different values of $\bar{u}/k$ are larger than the additive values.* For this reason, we relate ρ_m^* and ξ_m to V_{ij}^o, and we avoid any associations with the mixed collision diameters in Eq. (5.13) and, later, in Eq. (6.19) and the equations of state.

Other approaches to the statistics of chain molecules are considered by Pratt et al. (1978) and Hsu et al. (1978) and in earlier papers by Weeks, Chandler, and Anderson on the WCA perturbation theory. Fisher and Lago (1983) developed a theory, analogous to the WCA approach, for mixtures of one-center and two-center molecules. The agreement with saturated densities of pure fluids obtained from Monte-Carlo simulation is better than in the case of Barker-Henderson (BH) theory (Eq. 4.1). However, the excess functions of the systems Ar + N_2 and Ar + O_2 are predicted as well by the BH theory as by the more complicated WCA theories. This is not surprising because N_2 and O_2

(or CH_4) molecules are too small to be treated as two-center molecules (see the discussion of Table 8.1 and 8.2).

C. The Ordering Effects in Mixtures

The basic relations in the last chapter, (4.11), (4.15), and (4.33) to (4.36), were obtained with the assumption that the molecules of different species may freely replace each other within the volume of the system. They are the relations for random mixtures. There are two main reasons why a real system may be a non-random mixture or exhibit certain *ordering effects*.

The first is large differences between $\bar{u}_{jj}$ and $\bar{u}_{ii}$ of the components. This difficult problem is extensively discussed by Rowlinson (1969). A correction term, simple enough to be applicable in practice, was derived by Guggenheim and is considered at the end of Chap. 6. It appears that the correction is negligibly small. As mentioned in Chap. 4, Twu and Gubbins applied the relations for random mixtures to systems in which one of the components has strong multipole moments, and obtained excellent results. However, the ordering effects are functions of T and ρ^*, and tend to become very weak at densities equal to or less than the critical density. If this phenomenon is disregarded, the saturated densities of the liquid phase—calculated at pressures close to P^c—are larger than the observed values, and the liquid and vapor phases become identical at a pressure sometimes much higher than the observed P^c. A semiempirical correction term may be derived from Byers Brown's relation for the deviation δA^E (*see* Rowlinson, 1969):

$$\delta A^E / x_1 x_2 = \frac{1}{4} G_{ff}(\Delta f)^2 + \frac{1}{2} G_{fk}(\Delta f)(\Delta h) + \frac{1}{4} G_{kk}(\Delta h)^2 \quad (5.26)$$

where $\Delta f = (\bar{u}_{22} - \bar{u}_{11})/(\bar{u}_{11} + \bar{u}_{22})$; $\Delta h = (\sigma_{22}^3 - \sigma_{11}^3)/(\sigma_{11}^3 + \sigma_{22}^3)$; and G_{ff}, G_{fk}, and G_{kk} are certain fluctuation functions. Since the Lennard-Jones approximation, from which (5.26) was derived, exaggerates the effect of size differences, we may consider the first term only. By putting $\Delta f \approx \Delta T^c = (T_2^c - T_1^c)/(T_1^c + T_2^c)$, we may write the correction due to ordering effects as

$$A^{or}/RT = B_o(\Delta T^c)^2 x_i x_j f(T, \rho^*) \quad (5.27)$$

where B_o is a constant, and $f(T, \rho^*)$ is a function that we shall consider in Section B of Chap. 7. At the present time, we know only that A^{or} is negative.

Even if the molecules are hard spheres, the small spheres may tend to occupy the gaps between the large ones, but the omission of the two terms in (5.26) is probably justified in the case of mixtures of molecules of "intermediate" sizes ("volatile" liquids). However, when one or both of the components are long-chain molecules, δA^E due to the differences in molecular sizes is not negligible. Flory (1941) and Huggins (1941), by counting the number of ways of arranging N_r r-mer (polymer) molecules on available lattice sites and by filling up the remaining empty sites with N_1 monomer (solvent) molecules, obtained the famous relation for the excess entropy due to mixing molecules of different sizes, usually called the *combinatorial entropy*:

$$S^{EC}/R = x_1 \ln(x_1/\phi_1) + x_2 \ln(x_2/\phi_2) \tag{5.28}$$

where $\phi_1 = N_1/(N_1 + rN_r)$ and $\phi_2 = rN_r/(N_1 + rN_r)$ and are close to the volume fractions. Accordingly, we may put $r_1 = 1$, $r_2 = V_2^\circ/V_1^\circ$, $r_3 = V_3^\circ/V_1^\circ$, etc., for a multicomponent system.

As shown in the next chapter, the excess internal energy of mixing U^E of nonattracting hard spheres is very nearly zero. Also, if the monomer and the segments of the polymer have identical interaction energies, the mixture will be *athermal* (no heat effect will result from mixing). Hence the combinatorial part of the internal energy

$$U^{EC} = 0 \tag{5.29}$$

Since $A^{EC} = U^{EC} - TS^{EC}$, we obtain for the combinatorial free energy

$$A^{EC}/RT = x_1 \ln(\phi_1/x_1) + x_2 \ln(\phi_2/x_2) \tag{5.30}$$

For a multicomponent system,

$$A^{EC} = \sum_i x_i \ln(\phi_i/x_i) \tag{5.31}$$

The excess combinatorial chemical potential is then

$$\frac{\mu_i^{EC}}{RT} = \left[\frac{\partial(A^{EC}/RT)}{\partial N_i} \right]_{T, V_m, N_k(k \neq i)}$$
$$= \ln(\phi_i/x_i) + 1 - \phi_i/x_i \tag{5.32}$$

Eqs. (5.28) and (5.30) were later derived by Longuet-Higgins (1953) without using the lattice model. Another notable feature of this theory is that it starts with the Helmholtz energy and (5.30) is obtained first; the combinatorial entropy (5.28) follows by differentiation of (5.30) with respect to T; and (5.29) is the final result. However, the theory of Longuet-Higgins contains some intuitive arguments.

Refinements of the Flory-Huggins theory are outlined by Prigogine et al. (1957) and Guggenheim (1952).

The Flory-Huggins and the Longuet-Higgins theory are in principle not restricted to chain molecules and should therefore apply to aromatics and cycloalkanes as well. However, in these cases ϕ_i may be less close to the volume fractions.

Whereas A^{EC} is always negative, S^{EC} is always positive. The theory has been extensively tested against experimental results and proved to be at least qualitatively correct at high densities of the fluids. More details about these tests are reviewed by Tompa (1956). According to (5.30), A^{EC} is independent of ρ^*, but it should vanish for $\rho^* = 0$.

Large deviations from the combining rules for $\bar{u}_{ij}$, noted in the systems of molecules of different sizes, are greatly diminished when the Flory-Huggins A^{EC} is added to A_m^r calculated for a random mixture.

CHAPTER 6

Equations of State

An equation of state is a relation between P, V, T, and x_i. The simplest one is the equation for a mixture of perfect gases, (1.1). Any real gas conforms to Eq. (1.44) at $\rho_m \to 0$; however, higher terms of the expansion must be added at greater densities. Kirkwood (1935) and independently Mayer and Mayer (1940) have shown that

$$Z_m = P/\rho_m RT = 1 + \rho_m \sum_i \sum_j x_i x_j \beta_{ij} + \rho_m^2 \sum_i \sum_j \sum_k x_i x_j x_k \beta_{ijk}^{(3)}$$

$$+ 0(\rho_m^3) + \ldots \qquad (6.1)$$

where β_{ij} and $\beta_{ijk}^{(3)}$ are functions of T only, and $\beta_{ijk}^{(3)}$ is called the *third virial coefficient*. The function $\beta^{(3)}(T)$ is shown in Fig. 6.1. In the temperature range of practical importance, $\beta^{(3)}$ is positive and passes through a maximum. Here the differences in $\beta^{(3)}$ depending on the chemical nature of the fluid are the greatest. At high temperatures $\beta^{(3)}$ tends to a constant value that is practically the same for all substances. Extensive comparisons and calculations of $\beta^{(3)}$ were carried out by Sherwood and Prausnitz (1964) and by Chueh and Prausnitz (1967). According to the latter, the following approximation

$$\beta_{ijk}^{(3)} = (\beta_{ij}^{(3)} \beta_{jk}^{(3)} \beta_{ik}^{(3)})^{1/3} \qquad (6.2)$$

yields results in good agreement with experimental data for the systems with London energies. For a binary system, the triple sum in (6.1) reduces to ($i = j$ or $j = k$):

$$\beta_m^{(3)} = x_1^3 \beta_{111}^{(3)} + 3x_1^2 x_2 \beta_{112}^{(3)} + 3x_1 x_2^2 \beta_{122}^{(3)} + x_2^3 \beta_{222}^{(3)} \qquad (6.3)$$

The values of the virial coefficients obtained from PVT data in a given range of molar densities strongly depend on the number of terms used. Eq. (6.1) with estimated values of $\beta^{(3)}(T)$ should be used to obtain accurate values of $\beta(T)$, even if the data are limited to very low densities ($P < 1$ bar). Corrections due to the effect of $\beta^{(3)}(T)$ were included, for example, in the determinations of $\beta(T)$ of argon and krypton by Weir et al. (1967). In turn, to obtain accurate values of

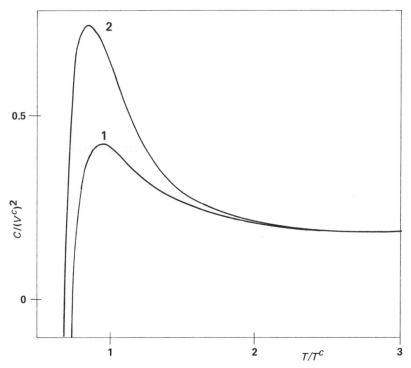

FIG. 6.1.   The third virial coefficients $\beta^{(3)}/(V^c)^2$ of argon and other small molecules (curve 1) and of *n*-octane (curve 2) as functions of T/T^c.

$\beta^{(3)}(T)$, the data must include *PVT* data at higher densities, and Eq. (6.1) must be expanded to higher terms with empirical coefficients. Errors due to improper truncation of this equation are discussed by Lee, Eubank, and Hall (1978) and by Holleran (1970).

If the virial equation were expanded to include the liquid state, an entirely impractical number of terms and combining rules would be required. Moreover, the third virial coefficient already contradicts the useful assumption of the additivity of pair interaction energies $u_{ij}(r)$. For these reasons the residual functions, derived for a real gas in Chap. 1, are limited to $\beta(T)$, and they are useful in the calculations of the solubilities of gases in liquids and the excess functions of mixtures at low pressures, up to about 5 bar.

The first equation of state with a theoretical foundation was derived by van der Waals (1873; see 1890). It is based on the considera-

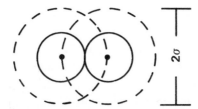

FIG. 6.2.   The centers of two hard spheres are separated by a volume equal to $4/3 \, \pi\sigma^3 = 8$ volumes of a sphere.

tion of the volumes occupied by a pair of hard spheres (the repulsion term) and the internal energy of liquids (the attraction term). The pair of spheres is shown in Fig. 6.2. Before the development of statistical mechanics, van der Waals foresaw that the probability of three-body collisions in a very dilute gas is practically nil. If only collisions between pairs are probable, and for each center a volume $4/3 \, \pi\sigma^3$ is excluded, then for $N/2$ pairs the volume is equal to $2/3 \, \pi N\sigma^3$, or four times the volume of N molecules. Consideration of the internal energy of liquids led van der Waals to the conclusion that it is a linear function of ρ; thus he arrived at the famous equation of state:

$$P = \frac{NkT}{V(1 - 4\xi)} - \frac{N^2 a}{V^2} \tag{6.4}$$

or

$$Z = \frac{1}{1 - 4\xi} - \frac{N_o^2 a \rho_m}{RT} \tag{6.5}$$

where a is the van der Waals "constant" and ξ is given by (4.24).

    With increasing density of the system, the probability of three-body, four-body (and so on) collisions increases until the random close-packed volume is reached, in which—according to (4.24)—only 0.26 of the total volume cannot be occupied. G. D. Scott (1960, 1962), by shaking 20,000 steel balls in metal cylinders, found that this volume equals $0.37 \, V$ ($\xi = 0.63$, $\rho^* = 0.85$). The numbers of nearest neighbors, or the so-called *coordination numbers* z as a function of ρ^*, are considered by Brostow and Sicotte (1975) and by Brostow (1977).

Near the triple point of argon, $z = 11$ ($\rho^* = 0.577$), and at the critical point $z = 5$ ($\rho^* = 0.218$). A solid phase may exist at $\rho^* > 0.735$.[†]

As mentioned earlier, Kac, Uhlenbeck, and Hemmer (1963) obtained the second term of the van der Waals equation as a result of very weak forces, $du(r)/dr \to 0$. Such conditions may exist when the molecules interact according to the square-well potential and are within the width of the well at all times: that is, in the liquid state. Hence, the second term of (6.5) is valid at high densities, the first at very low densities. Because of a partial cancellation of opposite errors, the van der Waals equation is qualitatively correct only at intermediate densities.

Many empirical equations of state have been proposed since then, and they usually contain the first term intact. The most popular equations used in the calculations of phase equilibria in mixtures are the Redlich-Kwong, the Peng-Robinson, and the Soave's equations. They are reviewed by K. C. Chao and Greenkorn (1975) and by Peng and Robinson (1976). As shown by Beret and Prausnitz on the example of Soave's equation, they must fail at high densities of a fluid. Nevertheless, the equations are used because they contain only two or three parameters.

Empirical equations of state, designed for accurate smoothing of the experimental data and construction of standard tables for pure substances, belong to a separate class. They contain up to 50 constants that are fitted to the data on Z, H^r, C_v, C_p, and $\beta(T)$. The most appropriate are the equations developed by Altunin and Gadetskii (1971), Wagner (1972), Jacobsen and Stewart (1973), Bender (1973), and Goodwin (1975). The Goodwin equation accounts for the unusual (nonanalytical) behavior of fluids in the critical region, whereas the others must be supplemented by relations valid in the critical region. Chapela and Rowlinson (1974) derived proper relations for this region, based on the scaling equations (Chap. 7) and certain switching functions to ensure a smooth passage to the region of validity of the Altunin-Gadetskii or the Bender equation. The application of all these tedious equations is best demonstrated by Angus, Armstrong, and de Reuck (1976) for carbon dioxide.

In the calculations of phase equilibria in mixtures, these equations

---

[†] The values of ρ^* were obtained from Eq. (6.26).

are important only when the principle of corresponding states is used, as outlined in Chap. 4. Here, an accurate equation of state is needed to evaluate $Z_o(T, V)$, $A_o^r(T, V)$, and $U_o^r(T, V)$ for the reference substance *only*.

The main problem of extending the first term of the van der Waals equation to high densities was solved nearly 90 years after the equation was first published. Reiss, Frisch, and Lebowitz (1959) developed the scaled-particle theory of fluids, which leads to the following equation of state for hard spheres:

$$Z^h = \frac{P^h}{\rho_n kT} = \frac{1 + \xi + \xi^2}{(1 - \xi)^3} \tag{6.6}$$

where superscript h indicates hard spheres (or hard convex bodies). The theory is outlined by Frisch (1964) and by Reiss (1967). At the same time, Percus and Yevick (1958) derived an integral equation in which $g(r_{12})\exp(u(r_{12})/kT)$ is related to $g(r_{13})$ and $g(r_{23})$. Wertheim (1963) and, independently, Thiele (1963) have solved this equation for a system of hard spheres and obtained the equation of state identical with (6.6). This theory is briefly reviewed by Henderson and Davison (1967) and by Barker and Henderson (1976).

As shown by Reiss (1967), the values of Z^h and the residual entropy S^{rh}/R calculated from (6.6) are in excellent agreement with the molecular dynamics simulation experiments of Alder and Wainwright (see Reiss), which extend up to $\rho^* = 0.68$.

Carnahan and Starling (1969) noted that conformity with the virial expansion for hard spheres requires one more term in (6.6); that is,

$$Z^h = \frac{1 + \xi + \xi^2 - \xi^3}{(1 - \xi)^3} = 1 + \frac{4\xi - 2\xi^2}{(1 - \xi)^3} \tag{6.7}$$

The properties of real fluids confirm that the Carnahan-Starling equation is better than (6.6). The equation proposed by Hall (1972) contains still more terms; it may be better than (6.7) for highly compressed liquids, and with one more term the Hall equation is valid for solids as well. Also, Speedy (1977) derived a more accurate equation for hard spheres; however, it is too complex to be used for mixtures.

B. D. Singh and Sinha (1977, 1978, 1979) derived the quantum corrections to the hard-sphere equation of state for very light molecules at low temperatures. The corrections are simple enough to be useful in practice.

Recently, Boublik (1975) derived the equation of state for convex hard bodies of any shape:

$$Z^h = \frac{1 + (3\alpha - 2)\xi + (3\alpha^2 - 3\alpha + 1)\xi^2 - \alpha^2\xi^3}{(1 - \xi)^3}$$

$$= 1 + \frac{(3\alpha + 1)\xi + (3\alpha^2 - 3\alpha - 2)\xi^2 + (1 - \alpha^2)\xi^3}{(1 - \xi)^3} \quad (6.8)$$

where $\alpha = sI_c\rho/3\xi$ and where ξ/ρ, s, and I_c stand, respectively, for the volume, the surface area, and the $(1/4\pi)$-multiple of the mean curvature integral of the hard convex body.[†] For hard spheres, $\alpha = 1$, and (6.8) simplifies itself to Eq. (6.7). Earlier, Gibbons (1969) derived an analogous equation with two terms in the numerator that for $\alpha = 1$ reduces to Eq. (6.6). Pavlicek, Nezbeda, and Boublik (1979) extended (6.8) to mixtures.

Beret and Prausnitz (1975) obtained, by approximate considerations of the partition function $Q = q^N/N!$, a different equation for nonspherical hard bodies, or rather hard chains. Following Prigogine, they separated q into $q_{int} \cdot q_{ext}$, where q_{int} depends only on T, but q_{ext} depends on T, ρ, and the total number of rotational and vibrational degrees of freedom, denoted by $3c$. For argon-like molecules or hard spheres, $c = 1$; for nonspherical molecules, $c > 1$. The residual Helmholtz energy, obtained by inserting (6.7) into (1.23) and by integration, is

$$\frac{A_m^{rh}(T, V)}{RT} = \frac{4\xi - 3\xi^2}{(1 - \xi)^2} \quad (6.9)$$

Beret and Prausnitz stated that the simplest form of q_{ext} satisfying all the boundary conditions ($\rho^* \to 0$, $\rho^* \to 1$, $c \to 1$) is

$$\ln q_{ext} = (1 - c) A_m^{rh}/RT \quad (6.10)$$

By using (1.8) and (1.22), one obtains the equation of state for hard chains:

[†]The geometry of convex bodies is treated in detail in the excellent book by Kihara (1976).

$$Z^h = 1 + \frac{c(4\xi - 2\xi^2)}{(1 - \xi)^3} \tag{6.11}$$

which for hard spheres ($c = 1$) becomes identical with (6.7).

The extension of the theory of hard spheres to mixtures of hard spheres is not a simple problem. It was solved by Lebowitz (1964) and Lebowitz and Rowlinson (1964) for Eq. (6.6), and extended to (6.7) by Mansoori and his colleagues (1971). The relations are

$$Z^h_m = [1 + (1 - 3z_1)\xi_m + (1 - 3z_2)\xi_m^2 - z_3\xi_m^3](1 - \xi_m)^{-3} \tag{6.12}$$

where

$$\xi_m = \frac{1}{6} \pi N_o \sum_i x_i \sigma_{ii}^3 / V_m \tag{6.13}$$

and the relations for the z's, due to Lebowitz, are

$$z_1 = \sum_{j > i = 1}^{m} \Delta_{ij}(\sigma_{ii} + \sigma_{jj})(\sigma_{ii}\sigma_{jj})^{-1/2}; \tag{6.14}$$

$$z_2 = \sum_{j > i = 1}^{m} \Delta_{ij} \sum_{k = 1}^{m} \left(\frac{\xi_k}{\xi_m}\right) \frac{(\sigma_{ii}\sigma_{jj})^{1/2}}{\sigma_{kk}}; \tag{6.15}$$

and

$$z_3 = \left[\sum_i \left(\frac{\xi_i}{\xi_m}\right)^{2/3} x_i^{1/3} \right]^3 \tag{6.16}$$

where

$$\Delta_{ij} = (\xi_i \xi_j x_i x_j)^{1/2}(\sigma_{ii} - \sigma_{jj})^2(\sigma_{ii}\sigma_{jj}\xi_m)^{-1} \tag{6.17}$$

For the pure components, $\Delta_{ij} = z_1 = z_2 = 0$; $z_3 = 1$.

Lebowitz and Rowlinson have shown that V^E and G^E of mixtures of hard spheres with different diameters are small and always negative. Mansoori et al. demonstrated a nearly exact agreement of these excess functions calculated from the foregoing relations with the molecular dynamics data of Alder (1964).

Testing the relations with data for real systems is less direct because it involves the intermolecular attraction term. If, following the discussion of the van der Waals equation, the second term in (6.5) is valid at liquid densities, the following equation should be valid at these densities:

$$Z = Z^h + Z^a = \frac{1 + \xi + \xi^2 - \xi^3}{(1 - \xi)^3} - \frac{N_o^2 a \rho_m}{RT} \tag{6.18}$$

where superscript a is for the attraction term. Longuet-Higgins and Widom (1964) confirmed by the evaluation of Z, U^r/RT, and S^r/RT that this equation is valid for liquid argon at and near the triple point. Snider and Herrington (1967) derived from (6.18) all the relations for the thermodynamic functions of mixtures. The parameter a_m was expressed by the van der Waals approximation (4.11). Snider and Herrington have shown for several liquids, including globular molecules such as cyclohexane, that if σ and ξ are calculated from the internal energy data, then the molar volumes calculated from (6.18) at a given T and P agree very well with the observed liquid volumes at temperatures up to at least the normal boiling point. These tests suffice to show that real molecules closely follow the behavior of hard spheres. The tests of Eq. (6.18) for mixtures made by Snider and Herrington, McGlashan (1970), Marsh, McGlashan, and Warr (1970), Miller (1971), and R. L. Scott (1972) are less convincing because, as shown by McGlashan, good results for the excess functions of systems of molecules *with similar sizes* are also obtained by using less accurate equations of state.

The applicability of Eq. (6.18) to mixtures of molecules of very different sizes was demonstrated by Kreglewski et al. (1973). The superiority of Lebowitz's relations (6.14) to (6.17) over Lorentz's rule (5.11) was clearly demonstrated here for the $n\text{-}C_6H_{14} + n\text{-}C_{16}H_{34}$ system, among others. Johnston and Eckert (1981) and Johnston et al. (1982) successfully applied Eq. (6.18) to the problem of solubility of high-molecular mass hydrocarbons in supercritical fluids.

As mentioned earlier, the simple rule (5.12) yields practically the same results as Lebowitz's relations. Therefore, Eqs. (6.7) and (6.8) may be extended directly to mixtures by putting

$$\xi = \xi_m = 0.74048 \, V_m^o \rho_m \tag{6.19}$$

where, in accordance with (4.12),

$$V_m^o = \sum_i \sum_j x_i x_j V_{ij}^o \tag{6.20}$$

As the temperature increases and a liquid approaches the critical point, the second term in (6.18) begins to fail. Ponce and Renon (1976) obtained a relatively simple relation for the attraction term of the Helmholtz energy A^{ra}/RT by neglecting the second integral in (4.1) and by expressing $u(r)$ by the square-well potential with a width of the well $1 < (r/\sigma) < s/\sigma$:

$$\frac{A_m^{ra}}{RT} = - \xi \frac{\bar{u}}{kT} \left[\frac{1 - \delta}{2\delta\xi} + 4\left(\frac{s}{\sigma}\right)^3 \right]$$

$$- 2\xi \left(\frac{\bar{u}}{kT}\right)^2 \frac{1}{\delta} \left[\left(\frac{s}{\sigma}\right)^3 - \left(\frac{\xi + 2}{2}\right)\left(\frac{1 - \xi}{2\xi + 1}\right)^3 \right] \tag{6.21}$$

where $\delta = (2\xi + 1)^2/(1 - \xi)^4$. Differentiation according to (1.22) gives

$$Z = Z^h - 4\xi \frac{\bar{u}}{kT} \left[\left(\frac{s}{\sigma}\right)^3 - \left(\frac{\xi + 2}{2}\right)\gamma \right]$$

$$+ 2\xi\gamma \left(\frac{\bar{u}}{kT}\right)^2 \left\{ \left[\left(\frac{s}{\sigma}\right)^3 - \left(\frac{\xi + 2}{2}\right)\gamma \right](6\xi^2 + 7\xi - 1) \right.$$

$$\left. - \frac{\xi\gamma}{2}(2\xi^2 + 8\xi + 17) \right\} \tag{6.22}$$

where $\gamma = [(1 - \xi)/(2\xi + 1)]^3$ and s/σ may be set equal to 3/2. Comparison of the calculated values of A^{ra}/RT with the results of molecular dynamics by Alder, Young, and Mark (1972) indicates that at high densities of a real fluid they will be less negative than the experimental values. Nevertheless, (6.21) certainly is a useful solution to the problem.

In the studies of molecular dynamics, Alder and his coworkers expressed A^{ra}/RT by a power series of ρ^*, and $\bar{u}/kT = 1/T^*$ as follows:

$$\frac{A_m^{ra}}{RT} = \sum_n \sum_m D_{nm}(1/T^*)^n(\rho^*)^m \tag{6.23}$$

where $\bar{u}/k$ is the minimum value of the square-well potential, assumed to have a constant $\mathfrak{R} = 3/2$, and D_{nm} are universal constants. Alder found it necessary to expand the series up to $n = 4$ and $m = 9$ with twenty-four D_{nm} constants to reproduce properly the results obtained up to the highest densities of a fluid. The intention of the power series is to replace the pair and the higher distribution functions, which cannot be evaluated directly. The universality of the D_{nm} constants means that the distribution functions of various fluids are identical functions of T^* and ρ^*. The authors called this treatment *the augmented van der Waals theory of fluids*. We shall use this name when (6.23) is combined with A_m^{rh}/RT derived from the theory of hard convex bodies.

Eq. (6.23) is able to reproduce accurately the internal energies and the heat capacities obtained by molecular dynamics at and in the vicinity of a critical point. As shown later, the results of comparisons with real fluids are satisfactory. The molecular dynamics results obtained for 108 and 500 atoms are compared with the data for argon in Table 6.1.

According to the new theory of nonspherical molecules developed by Chandler and Pratt (1976) and by Pratt and Chandler (1977), the radial distribution functions are different from those of spherical molecules. However, with the exception of strongly polar and hydrogen-bonded fluids, experimental *PVT* data of various fluids do not disclose any systematic deviations from A_m^{ra} given by (6.23) with the constants D_{nm} fitted to the data for argon or ethane. It seems that A^r of real fluids is not very sensitive to the form of $g(r)$ but rather to $g(T^*, \rho^*)$.

Beret and Prausnitz (1975) combined (6.11) with Z^a derived from (6.23), obtaining

$$Z = 1 + \frac{c(4\xi - 2\xi^2)}{(1 - \xi)^3} + \sum_n \sum_m m\, D_{nm}(1/cT^*)^n(\rho^*)^m \qquad (6.24)$$

TABLE 6.1. A comparison of critical constants obtained by molecular dynamics (SW potential) with those of argon.

	V^c/V^o	$kT^c/\bar{u}$	$P^cV^c/RT^c = Z^c$
108 atoms	4.40 ± 0.2	1.290 ± 0.005	0.326
500 atoms	4.25 ± 0.25	1.260 ± 0.005	0.287
Argon	4.58 ± 0.05	1.261 [a]	0.2913

(a) Beret and Prausnitz (1975).

The second virial coefficient, $\beta = \lim_{\rho \to 0}(\partial Z/\partial\rho)_T$, derived from (6.24) is

$$\frac{\beta}{cV^o} = 2.962 + \frac{D_{11}}{cT^*} + \frac{D_{21}}{(cT^*)^2} + \frac{D_{31}}{(cT^*)^3} + \frac{D_{41}}{(cT^*)^4} \quad (6.25)$$

where $2.962 = 4 \times 0.74048$. Hence, the limiting value for nonspherical molecules as $T \to \infty$ is c times greater than $2/3 \, \pi N_o \sigma^3$. This may mean rather that β tends to $2/3 \, \pi N_o(\sigma^e)^3$ where the effective diameter in the gas phase is $c^{1/3}$ times greater than the diameter at high densities. The authors obtained good results for the compressibilities and the enthalpies of gases and liquids and for the saturated vapor pressures. Donohue, Kaul, and Prausnitz (1976, 1977) extended this equation of state to mixtures and demonstrated its accuracy in the calculations of liquid-vapor equilibria and gas solubilities in alkane + alkane or CO_2 or H_2S systems. The details of the method were not published.

It is now well known that if an equation of state holds well for pure fluids, it will also be good for mixtures—but not vice versa. For example, the Ponce-Renon equation (6.21) was compared only with molecular dynamics data for pure fluids, and the good results indicate that it is one of the reliable equations for mixtures.

S. S. Chen and Kreglewski (1977) combined Boublik's equation (6.8) with Z^a derived from (6.23) as follows:

$$Z_m = Z_m^h + Z_m^a$$

$$= 1 + \frac{(3\alpha_m + 1)\xi_m + (3\alpha_m^2 - 3\sigma_m - 2)\xi_m^2 + (1 - \alpha_m^2)\xi_m^3}{(1 - \xi_m)^3}$$

$$+ \sum_n \sum_m m \, D_{nm}(1/T_m^*)^n (\rho_m^*)^m \quad (6.26)$$

where $1/T_m^* = \bar{u}_m/kT$ and the subscript m indicates that the equation is applied to mixtures as well as to pure fluids. In $\rho_m^* = V_m^o/V_m$ and $\xi_m = 0.74048\rho_m^*$, V_m^o is given by (6.20), and $\bar{u}_m/k$ is expressed by the square-well approximation (4.15); but the van der Waals approximation can be used alternatively, since the two become identical for the pure components. In the exact treatment (Boublik, 1975; Boublik and Lu, 1978), Z_m^h, A_m^{rh} of mixtures and μ_i^{rh} are more complicated functions of s, I_c and ξ/ρ in $\alpha = sI_c\rho/3\xi$, which are averaged separately. The values of α obtained from *PVT* data of fluids are always close to unity, and the simplified Eq. (6.26) is as good as the exact relations. For the same reason we assume that

$$\alpha_m = \sum_i x_i \sigma_i \tag{6.27}$$

Possible deviations from (6.27) are felt even less than those from (6.20).

The functions $Y^r(T, V)$ obtained from (6.26) by applying the basic relations (1.23) and (1.24), respectively, are

$$\frac{A_m^r}{RT} = \frac{A_m^{rh} + A_m^{ra}}{RT}$$

$$= (\alpha_m^2 - 1)\ln(1 - \xi_m) + \frac{(\alpha_m^2 + 3\alpha_m)\xi_m - 3\alpha_m\xi_m^2}{(1 - \xi_m)^2}$$

$$+ \sum_n \sum_m D_{nm}(1/T_m^*)^n(\rho_m^*)^m \tag{6.28}$$

and

$$\frac{U_m^r}{RT} = \frac{U_m^{rh} + U_m^{ra}}{RT} = \frac{T}{\xi_m}\left(\frac{\partial \xi_m}{\partial T}\right)_v (1 - Z_m)$$

$$+ F(\eta)\sum_n \sum_m n \, D_{nm}(1/T_m^*)^n(\rho_m^*)^m \tag{6.29}$$

where

$$F(\eta) = -\frac{kT^2}{\bar{u}_m}\left[\frac{\partial(\bar{u}_m/kT)}{\partial T}\right]_v = -\frac{kT^2}{\varepsilon_m}\left[\frac{\partial(\varepsilon_m/kT)}{\partial T}\right]_v \tag{6.30a}$$

Differentiation at constant volume is indicated in (6.30) because $\bar{u}_m/k$ may depend on ρ^* for large molecules. From Eqs. (5.8), (5.10), and (5.23), we obtain

$$F(\eta) = +\frac{kT^2}{\varepsilon_m}\sum_i \sum_j \varphi_i \varphi_j \, \varepsilon_{ij}^o\left[1 + \frac{2\eta_{ij}}{kT} + 3\left(\frac{\delta_{ij}}{kT}\right)^2\right] \tag{6.30b}$$

Hence, the effect of η and δ is much stronger in U^r than in A^r; therefore, it is also stronger in H^E than in G^E of mixtures. Further, by using (1.25), we obtain

$$\frac{C_{vm}^r}{R} = -\left[2T\left(\frac{\partial \xi_m}{\partial T}\right)_v + T^2\left(\frac{\partial^2 \xi_m}{\partial T^2}\right)_v\right]\frac{(Z_m^h - 1)}{\xi_m}$$

$$- T^2\left(\frac{\partial \xi_m}{\partial T}\right)_v^2\left[\frac{4\alpha_m^2 + 6\alpha_m + (2\alpha_m^2 - 6\alpha_m)\xi_m}{(1 - \xi_m)^4} - \frac{(\alpha_m^2 - 1)}{(1 - \xi_m)^2}\right]$$

$$- Z_m^a\left[\frac{2T}{V_m^o}\left(\frac{\partial V_m^o}{\partial T}\right)_v - \left(\frac{T}{V_m^o}\right)^2\left(\frac{\partial V_m^o}{\partial T}\right)_v^2 + \frac{T^2}{V_m^o}\left(\frac{\partial^2 V_m^o}{\partial T^2}\right)_v\right]$$

$$+ T\frac{\partial F(\eta)}{\partial T}\sum_n\sum_m n\, D_{nm}(1/T_m^*)^n(\rho_m^*)^m$$

$$- F(\eta)\sum_n\sum_m n(n - 1)\, D_{nm}(1/T_m^*)^n(\rho_m^*)^m$$

$$+ \frac{T}{V_m^o}\left(\frac{\partial V_m^o}{\partial T}\right)_v\left[2F(\eta)\sum_n\sum_m nm\, D_{nm}(1/T_m^*)^n(\rho_m^*)^m\right.$$

$$\left. - \frac{T}{V_m^o}\left(\frac{\partial V_m^o}{\partial T}\right)_v \sum_n\sum_m m^2\, D_{nm}(1/T_m^*)^n(\rho_m^*)^m\right] \tag{6.31}$$

The functions $Y^r(T, P)$, derived in Chap. 1, are

$$\frac{G_m^r}{RT} = \frac{A_m^r(T, V)}{RT} + Z_m - 1 - \ln Z_m \tag{6.32}$$

and

$$\frac{H_m^r}{RT} = \frac{U_m^r}{RT} + Z_m - 1 \tag{6.33}$$

The residual heat capacity C_p^r is given by (1.32), where

$$\left(\frac{\partial P}{\partial T}\right)_v = \frac{R}{V_m}\left\{Z_m\right.$$

$$+ T\left(\frac{\partial \xi_m}{\partial T}\right)_v\frac{[1 + 3\alpha_m + (6\alpha_m^2 - 2)\xi_m - (3\alpha_m - 1)\xi_m^2]}{(1 - \xi_m)^4}$$

$$- F(\eta)\sum_n\sum_m nm\, D_{nm}(1/T_m^*)^n(\rho_m^*)^m$$

$$\left. + \frac{T}{V_m^o}\left(\frac{\partial V_m^o}{\partial T}\right)_v\sum_n\sum_m m^2\, D_{nm}(1/T_m^*)^n(\rho_m^*)^m\right\} \tag{6.34}$$

and

$$\left(\frac{\partial P}{\partial V_m}\right)_T =$$

$$-\frac{RT}{V_m^2}\left[\frac{1 + (6\alpha_m - 2)\xi_m + (9\alpha_m^2 - 6\alpha_m + 1)\xi_m^2 - 4\alpha_m^2\xi_m^3 + \alpha_m^2\xi_m^4}{(1 - \xi_m)^4}\right.$$

$$\left. + Z_m^a + \sum_n\sum_m m^2\, D_{nm}(1/T_m^*)^n(\rho_m^*)^m\right] \tag{6.35}$$

The second virial coefficient is

$$\beta_m/V_m^o = 0.74048(1 + 3\alpha_m) + \sum_{n=1}^{4} D_{n1}(1/T_m^*)^n \tag{6.36}$$

if the power series is expanded up to $n = 4$.

The residual chemical potential of the component i is given by (1.38), where in the case of the equation of state (6.26),

$$\left[\frac{\partial(A_m^r/RT)}{\partial x_j}\right]_{T,\, V,\, x_{k(k \neq i,\, j)}} = \left[2\alpha_m \ln(1 - \xi_m)\right.$$

$$+ \frac{(2\alpha_m + 3)\xi_m - 3\xi_m^2}{(1 - \xi_m)^2}\right]\left(\frac{\partial\alpha_m}{\partial x_j}\right)_{x_k}$$

$$+ \frac{1}{\xi_m}\left(\frac{\partial\xi_m}{\partial x_j}\right)_{T,\, V,\, x_k}(Z_m - 1)$$

$$+ T_m^*\left[\frac{\partial(\bar{u}_m/kT)}{\partial\varphi_j}\frac{\partial\varphi_j}{\partial x_j}\right]_{x_k}\sum_n\sum_m n\, D_{nm}(1/T_m^*)^n(\rho_m^*)^m \tag{6.37}$$

where

$$\left(\frac{\partial\xi_m}{\partial x_j}\right)_{T,\, V,\, x_k} = 0.74048\, \rho_m\left(\frac{\partial V_m^o}{\partial x_j}\right)_{x_k} \tag{6.38}$$

If $\bar{u}_m$ is given by the square-well approximation, $(\partial\bar{u}_m/\partial\varphi_i)_{x_k}$ are obtained from (1.53), with x_i replaced by φ_i. The derivatives $\partial\varphi_i/\partial x_i$ in (6.37) are

$$\left(\frac{\partial \varphi_i}{\partial x_i}\right)_{x_k} = \frac{\varphi_i}{x_i}(1 - \varphi_i) + \frac{\varphi_i \varphi_j}{x_j} ; \qquad (k \neq i, j) \qquad (6.39a)$$

When all the mole fractions are variable, then

$$\frac{\partial \varphi_i}{\partial x_i} = \frac{\varphi_i}{x_i}(1 - \varphi_i) + \varphi_i \sum_{j \neq i}\left(\frac{\varphi_j}{x_j}\right) \qquad (6.39b)$$

The relations corresponding to the Beret-Prausnitz equation of state are analogous to Eqs. (6.28) through (6.35) and (6.37), with $\alpha_m = 1$ and appropriately replaced by c, which for mixtures—according to Flory et al. (1964)—is additive:

$$c_m = \sum_i \phi_i c_i \qquad (6.40)$$

where ϕ_i are the site fractions.

The derivatives $(\partial \xi/\partial T)_v$ and $(\partial V^o/\partial T)_v$ are functions of the temperature dependence of the collision diameters. They are zero for hard spheres. However, real molecules behave rather like elastic spheres or interpenetrable spheres. Already, Reiss and coworkers have noted in comparisons with internal energies of real fluids that the diameters in the scaled-particle theory must depend on T. Wilhelm (1973, 1974) and others mentioned in his papers made attempts to determine this dependence. The most convincing results are those based on C_v^r/R of liquid argon, presented in Fig. 6.3. The curve shows the values derived from the Carnahan-Starling equation (6.7), assuming that σ is a linear function of T, $(d\sigma/dT)/\sigma = -0.16 \cdot 10^{-3} \mathrm{K}^{-1}$ and that σ at the triple point of argon equals 3.48 Å. The use of a more realistic equation of state—say, (6.18)—in which $a \sim \bar{u}\sigma^3$ would lead to a better estimate of $d\sigma/dT$.

Barker and Henderson (1967) have shown that the effective collision diameter is given by the relation

$$\sigma(T) = \int_0^{\sigma}[1 - \exp(-u(r)/kT)]dr \qquad (6.41)$$

Since in the range $0 < r < \sigma$, $u(r)$ is positive, $\sigma(T)$ decreases with rising T. In the limit $u(r) = \infty$ for hard spheres, σ is independent of T. If $u(r)$ in this range is expressed by an "inverted" square well, as

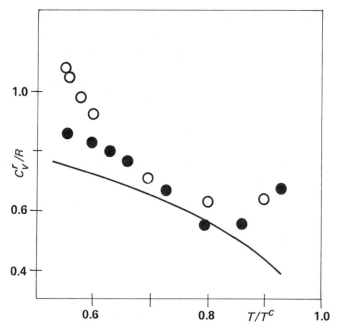

FIG. 6.3.   The residual heat capacity C_v^r/R of saturated liquid argon as a function of the reduced temperature T/T^c. The experimental values are according to Kohler (1972) and Rowlinson (1969), open circles and filled circles, respectively. The line is calculated by Wilhelm as explained in the text.

shown in Fig. 3.5 (first step), the integration according to (6.41) gives the result

$$\sigma(T) = \sigma^o[1 - C \cdot \exp(-3\bar{u}^o/kT)]$$

or

$$V^o = V^{oo}[1 - C \cdot \exp(-3\bar{u}^o/kT)]^3 \qquad (6.42)$$

where $C = s_1/\sigma$, and σ^o or V^{oo} are the values at $T = 0$. It is assumed arbitrarily that $u(r)$ in this range is equal to $3\bar{u}^o$. Kreglewski and Chen (1975) evaluated C from compressibility data for seven gases and found that C slightly decreases with increasing size of the molecules. However, by considering only the most accurate data, C appears to be a constant, with a value of $C \approx 0.12$. This is an average value over the temperature ranges of the data, below $T/T^c \approx 3$, and at higher temperatures C may be larger. The data for $\beta(T)$ indicate that C is also larger for water and other hydrogen-bonded gases. The calculated values of C_v^r agree well with the data for liquid argon down to $T/T^c \approx 0.7$. Below this temperature, $\sigma(T)$ apparently varies with T faster than Eq. (6.42) suggests. It is important, however, that this equation is simple and contains only one constant, which is nearly the same for most fluids. The derivatives $\partial \xi / \partial T$, and so on, that appear in (6.29) through (6.34) are

$$\frac{T}{\xi_m} \left(\frac{\partial \xi_m}{\partial T} \right)_v = \frac{T}{V_m^o} \left(\frac{\partial V_m^o}{\partial T} \right)_v$$

$$= -\frac{9C\bar{u}^o}{kT} [\exp(+3\bar{u}^o/kT) - C]^{-1} \tag{6.43}$$

and

$$\left(\frac{\partial^2 \xi_m}{\partial T^2} \right)_v = \left(\frac{\partial^2 V_m^o}{\partial T^2} \right)_v = \frac{1}{T} \left(\frac{\partial \xi_m}{\partial T} \right)_v \left\{ \frac{3\bar{u}^o}{kT} \left[1 \right. \right.$$

$$\left. \left. - \frac{2C}{\exp(+3\bar{u}^o/kT) - C} \right] - 2 \right\} \tag{6.44}$$

The values of the universal constants D_{nm} are a separate problem. Expansion of the square-well potential into a power series gives, for $\alpha = 1$ and $\mathcal{R} = 3/2$,

$$\frac{\beta}{V^o} = 2.962 - 7.0346 \left[\frac{\bar{u}}{kT} \right.$$

$$\left. + \frac{1}{2} \left(\frac{\bar{u}}{kT} \right)^2 + \frac{1}{6} \left(\frac{\bar{u}}{kT} \right)^3 + \frac{1}{24} \left(\frac{\bar{u}}{kT} \right)^4 + \ldots \right] \tag{6.45}$$

Hence, the constants D_{11}, D_{21}, D_{31}, and D_{41} in (6.25) or (6.36) should be negative. Alder et al. (1972) used the four constants resulting from (6.45) and fitted the remaining twenty constants to the molecular dynamics results. Donohue and Prausnitz (1976) found that these constants do not reproduce the data for real fluids well enough and proposed a new set of twenty-one constants, four of them being those

TABLE 6.2.   The universal constants D_{nm} according to Donohue and Prausnitz (1976).

$D_{11} = -7.0346$ [a]	$D_{31} = -1.17$ [a]
$D_{12} = -7.2736$	$D_{32} = +7.15$
$D_{13} = -1.252$	$D_{33} = -31.3$
$D_{14} = +6.0825$	$D_{34} = +63.107$
$D_{15} = +6.8$	$D_{35} = -40.608$
$D_{16} = +1.7$	
$D_{21} = -3.52$ [a]	$D_{41} = -0.29$ [a]
$D_{22} = +11.15$	$D_{42} = -1.32$
$D_{23} = -10.69$	$D_{43} = +32.9$
$D_{24} = -3.598$	$D_{44} = -94.248$
$D_{25} = +7.432$	$D_{45} = +73.387$

(a) The coefficients of Eq. (6.45).

from (6.45). The constants are given in Table 6.2. Again, these constants do not reproduce the second virial coefficients well, and Donohue proposed an empirical correction term in (6.25).

Kreglewski and Chen (1975) investigated whether this failure is due to an inherent inaccuracy of the square-well potential. All the D_{nm} constants were obtained by using $\beta(T)$, Z, and U^r/RT data for argon, accurately smoothed by Angus and Armstrong (1971), as follows. The molecular dynamics results have the advantage that they are obtained as functions of V^o/V, and V^o need not be known to obtain the constants D_{nm}, whereas when the data for argon are used for this purpose, V^o must be known in advance. Bienkowski and Chao (1975) assumed that at $P \to \infty$ the coefficient of isothermal compressibility $-(1/V)(\partial V/\partial P)_T$ depends almost exclusively on the hard-sphere term. Therefore, they extrapolated the "reduced" values $\xi^c = 1/6\ \pi N_o \sigma^3/V^c$ obtained from the data for argon, krypton, and xenon to $P = \infty$. The values of σ so obtained vary approximately linearly with T/T^c:

$$(\xi^c)^{1/3} = 0.54906 - 0.010827\ T/T^c \tag{6.46}$$

Hence, at $T = 0$, $\xi^c = 0.1655$, and for argon ($V^c = 74.59$ cm³mol⁻¹) $V^{oo} = 16.67$ cm³mol⁻¹. Kreglewski and Chen found that the *PVT* data for argon and other gases at $T/T^c > 4/3$ are very well reproduced by the simple equation of state (6.18), with the parameter a expressed by $a = a^o(1 + 0.1223\ \rho/\rho^c)$, where a^o is a characteristic constant, and

V^o is given by (6.42). The constants a^o, C, and V^{oo} were obtained by least squares with the result $V^{oo} = 16.29$ cm^{3}mol^{-1} for argon. This value is probably not more uncertain than that reported by Bienkowski and Chao.

Since the data in Table 3.4 indicate that $\bar{u}/k$ of argon is close to T^c, it is proper to assume $\bar{u}/k = T^c = 150.86$ K. By using the values $\alpha = 1$, $V^{oo} = 16.29$ cm^{3}mol^{-1}, $C = 0.12$ (in Eq. 6.42b), and $\bar{u}/k = 150.86$ K, the following values were obtained from $\beta(T)$ data for argon:[†]

$$D_{11} = -8.8043; \quad D_{21} = +2.9396; \quad D_{31} = -2.8225;$$
$$D_{41} = +0.3400 \tag{6.47}$$

The errors in the values of $\beta(T)$ calculated by using the coefficients of (6.45) are not acceptable; therefore, the empirical coefficients (6.47) were adopted, although the power series is then divorced from the square-well potential. The next twenty D_{nm} constants were obtained by S. S. Chen and Kreglewski (1977) from the data for liquid and gaseous argon and are given in Table 6.3(A). They are not valid below $T/T^c \approx 0.55$, corresponding to the triple point of argon. Since the excess functions of many systems were determined at lower reduced temperatures, Kreglewski and Chen (1978) proposed an *alternative* set of sixteen D_{nm} constants fitted to the data for liquid and gaseous ethane (which is liquid down to $T/T^c = 0.296$), given by Goodwin, Roder, and Straty (1976). The results obtained with these constants near to T^c at densities $\rho > \rho^c$ were unsatisfactory, and the author revised them recently and increased the number of the constants to eighteen, plus the set given in (6.47). These values are presented in Table 6.3(B).

The values of the constants D_{nm} depend strongly on the method of weighing the experimental data. The weights of the internal energy data points were taken as equal to $10^3 \cdot U^r/RT$, and the weights of the compressibility data to ρ_m^2. The agreement with the critical point was enforced by giving this point a very large weight. However, the agreement with the conditions of a critical point, $(\partial P/\partial \rho)_T = (\partial^2 P/\partial \rho^2)_T = 0$, was not enforced, because it reduces the accuracy of the equation in another area.

The characteristic constants in (6.24) are c, V^o, and $\bar{u}/k$. Beret

---

[†] For the square-well potential, $\bar{u}/k = T^c/1.261$. The assumption that $\bar{u}/k = T^c$ affects the values but not the signs of the constants D_{nm}.

TABLE 6.3.   The universal constants D_{nm} in the augmented van der Waals theory of fluids.

(A) Primary set Reference fluid: argon		(B) Secondary set Reference fluid: ethane	
$D_{11} =$	-8.8043	$D_{11} =$	-8.8043
$D_{12} =$	$+4.1646270$	$D_{12} =$	$+31.42290$
$D_{13} =$	-48.203555	$D_{13} =$	-212.8219
$D_{14} =$	$+140.43620$	$D_{14} =$	$+566.7182$
$D_{15} =$	-195.23339	$D_{15} =$	-690.4289
$D_{16} =$	$+113.51500$	$D_{16} =$	$+328.1323$
$D_{21} =$	$+2.9396$	$D_{21} =$	$+2.9396$
$D_{22} =$	-6.0865383	$D_{22} =$	-17.90504
$D_{23} =$	$+40.137956$	$D_{23} =$	-55.76158
$D_{24} =$	-76.230797	$D_{24} =$	$+1193.777$
$D_{25} =$	-133.70055	$D_{25} =$	-5626.094
$D_{26} =$	$+860.25349$	$D_{26} =$	$+12909.05$
$D_{27} =$	-1535.3224	$D_{27} =$	-16140.81
$D_{28} =$	$+1221.4261$	$D_{28} =$	$+10618.08$
$D_{29} =$	-409.10539	$D_{29} =$	-2899.506
$D_{31} =$	-2.8225	$D_{31} =$	-2.8225
$D_{32} =$	$+4.7600148$	$D_{32} =$	$+9.372313$
$D_{33} =$	$+11.257177$	$D_{33} =$	-10.95458
$D_{34} =$	-66.382743	$D_{34} =$	$+4.926320$
$D_{35} =$	$+69.248785$	$D_{35} =$	0
$D_{41} =$	$+0.3400$	$D_{41} =$	$+0.3400$
$D_{42} =$	-3.1875014	$D_{42} =$	-0.8831261
$D_{43} =$	$+12.231796$	$D_{43} =$	$+0.5599758$
$D_{44} =$	-12.110681	$D_{44} =$	0

NOTE: The constants D_{nm} should not be rounded off further, because the values of P or Z at low temperatures and high densities are sensitive to the last digits. The arithmetic of double precision (16 digits) is required.

and Prausnitz assumed $\bar{u}/k$ to be independent of the temperature, whereas the temperature dependence of V^o was expressed by an empirical relation for methane, assumed to be valid also for other fluids.

The characteristic constants in (6.26) are α, V^{oo}, $\bar{u}^o/k$, and η/k (the constant $C = 0.12$, except in special cases). In accordance with the theory of noncentral forces, η/kT^c for systems with London energies is proportional to ω:

$$\eta^L/kT^c = 0.505\omega + 0.702\omega^2; \qquad (\pm 5\%) \qquad\qquad (6.48)$$

The coefficients are empirical. The relation is more accurate than that suggested earlier, $\eta^L/k \approx 0.60\omega T^c$, from which η/k was estimated for some of the substances in Table 6.4.

The characteristic constants of Eq. (6.26) are given in Table 6.4. Unless otherwise indicated, the constants are based on extensive *PVT* and U^r/RT data for liquids and gases. References to the experimental data are given by S. S. Chen and Kreglewski (1977), Zwolinski et al. (tables of the *j* series, since 1976), and Simnick, Lin, and Chao (1979). The latter authors enforced the thermodynamic conditions of the critical point (substances marked by superscript *a*). For these substances, the densities of the coexisting phases meet at temperatures slightly below the experimental values of T^c. For most of the remaining substances, the condition that the densities of the coexisting phases and the slopes $(\partial\rho'/\partial T)_c = (\partial\rho''/\partial T)_c$ become identical at the observed T^c, P^c, and ρ^c was enforced. In these cases, the condition $(\partial P/\partial\rho)_T = (\partial^2 P/\partial\rho^2)_T = 0$ is fulfilled slightly above the observed values of T^c. The above conditions can be enforced as follows: we add four fictitious points around the observed point (P^c, V^c, T^c) differing only in volumes, by ± 1 percent and ± 2 percent. For example, if $V^c = 100$ cm^{3}mol^{-1}, we add points $V = 98, 99, 101,$ and 102 cm^{3}mol^{-1}, where all five points have identical $P = P^c$ and $T = T^c$. We ascribe a very high weight to these five points compared with the weights of the remaining data points used for determination of the constants. The coexistence curve, calculated from Gibbs equilibrium conditions, agrees then very well with the observed values, including the critical region (Fig. 6.5). Exceptionally, the constants of hydrogen and helium were determined from gas compressibility data far above the critical point. For helium, two sets of constants are given in Table 6.4: (1) from gas compressibility data and (2) $\bar{u}/k = 10.5$ K from Table 5.1, and $V^{oo} = 8.14$ cm^{3}mol^{-1}, calculated from the data quoted in connection with Table 5.2, where $\sigma^3_{Ar}/\sigma^3_{He} = V^{oo}_{Ar}/V^{oo}_{He} = 2.002$ and $V^{oo}_{Ar} = 16.29$. The value of α for xenon seems to be too large and consequently $\bar{u}^o/k$ is higher than the critical temperature (289.73 K). The experimental data for xenon are less accurate than those for other simple molecules. The equation of state for water and the alcohols is valid only at relatively high reduced temperatures. The range of validity could be extended to lower temperatures by including the term $\Delta_{ii}/(kT)^2$ of Eq. (5.10a).

TABLE 6.4. Characteristic constants of Eq. (6.26) obtained for η/k given by Eq. (6.48) and the D_{nm} constants for argon (Table 6.2)

	α	$\dfrac{V^{oo}}{\dfrac{cm^3}{mol}}$	C	$\dfrac{\bar{u}^o/k}{K}$	$\dfrac{\eta/k}{K}$
Hydrogen ($T > 100$ K)[a]	1.0004	13.625	0.241	39.171	0.499[d]
Helium-4 ($T > 50$ K) (1)	1.	15.27	0.39	16.77	0
(2)	1.	8.14	0.39	10.5	0
Argon	1.	16.29	0.12	150.86	0
Xenon[c]	1.0231	25.499	0.12	294.38	1.2
Nitrogen	1.000[b]	19.457	0.12	123.53	3.0
Water ($T > 500$ K)	1.1432	11.010	0.172	586.20	180.[d]
Carbon monoxide[c]	1.0153	19.820	0.12	130.46	4.2
Carbon dioxide	1.0571	19.703	0.12	284.28	40.[d]
Sulfur dioxide[c]	1.0710	25.346	0.12	383.56	88.[d]
Hydrogen sulfide	1.044	20.672	0.12	373.66	15.[d]
Methane	1.000[b]	21.576	0.12	190.29	1.
Ethane	1.037	31.118	0.12	298.03	19.[d]
Propane	1.041	42.598	0.12	353.11	34.
n-Butane	1.051	53.855	0.12	399.56	51.
iso-Butane[c]	1.0482	54.682	0.12	383.11	47.
n-Pentane[a]	1.0566	65.751	0.12	435.83	70.72
iso-Pentane[a]	1.0565	64.958	0.12	432.20	62.71
Neopentane[a]	1.0498	65.518	0.12	409.59	51.28
n-Hexane[a]	1.0720	77.228	0.12	468.33	90.11
n-Heptane	1.0626	90.404	0.12	465.99	130.[d]
n-Octane[a]	1.0981	96.556	0.12	517.52	134.5
n-Decane[a]	1.1349	110.72	0.12	558.07	181.57
2,2,4-Trimethylpentane	1.0588	101.27	0.12	468.62	125.
Ethylene	1.0330	28.031	0.12	279.55	12.
Propylene[c]	1.0443	38.979	0.12	350.07	32.
1-Butene	1.0539	50.224	0.12	393.38	52.[d]
Methylpropene[c]	1.0544	50.140	0.12	394.54	49.
Cyclohexane[a]	1.0583	64.772	0.12	522.46	70.72
Benzene[c]	1.0613	54.289	0.12	529.24	72.
Toluene	1.0944	65.843	0.12	558.94	91.
o-Xylene	1.0731	77.902	0.12	549.56	147.[d]
m-Xylene	1.0820	78.725	0.12	538.67	150.[d]
p-Xylene	1.0808	79.403	0.12	537.83	149.[d]
Naphthalene[c]	1.0825	83.301	0.12	687.10	136.
Methanol ($T > 360$ K)[c]	1.1303	25.102	0.249	380.66	280.[d]

TABLE 6.4 (cont'd.)

	α	$\dfrac{V^{oo}}{\dfrac{cm^3}{mol}}$	C	$\dfrac{\tilde{u}^o/k}{K}$	$\dfrac{\eta/k}{K}$
Ethanol ($T > 300$ K)	1.1165	36.560	0.241	357.37	317.
1-Propanol ($T > 320$ K)[c]	1.0998	47.810	0.242	373.23	316.[d]
iso-Propanol					
($T > 320$ K)[c]	1.0998	47.799	0.169	341.53	328.
Acetone	1.1115	43.761	0.0453	467.15	111.
Diethyl ether[c]	1.0686	59.728	0.12	416.60	93.
t-Butyl-methyl ether[c]	1.0587	69.077	0.12	453.58	80.
Furan[c]	1.0648	41.788	0.12	464.69	61.
Tetrahydrofuran[c]	1.0687	48.132	0.12	512.09	70.
Quinoline[c]	1.0635	79.164	0.12	698.54	150.[d]
Iso-quinoline[c]	1.0409	79.704	0.12	704.66	150.[d]
Methanethiol[c]	1.0559	31.074	0.12	454.83	43.
Ethanethiol[c]	1.0619	42.928	0.12	474.19	59.
Thiophene[c]	1.0559	48.581	0.12	543.87	73.
Tetrahydrothiophene[c]	1.0678	55.637	0.12	600.55	80.
Perfluoro-methane[c]	1.0338	30.698	0.12	209.03	28.[d]

(a) Constants determined by Simnick, Lin, and Chao (1979).
(b) Assumed value of α. The value obtained by the method of least squares from *PVT* data is slightly less than unity.
(c) Constants determined by using only the liquid-vapor equilibrium data.
(d) η/k estimated by fitting enthalpy data. Eq. (6.48) yields, for example, $\eta/k = 114$ for *n*-heptane, 71 for SO_2, 77 for CO_2, 135 for water, and 171 K for methanol.

The results reviewed below demonstrate that the characteristic constants in (6.24) and (6.26) are not just empirical parameters but have a clear physical meaning. One of the interesting results, shown by Simnick, Lin, and Chao, is that ($\alpha - 1$) is approximately a linear function of the acentric factor. For substances with London energies and with less than seven carbon atoms,

$$\alpha - 1 \approx 0.25\omega \qquad (6.49)$$

In any case, α is much smaller than the values calculated by Ewing and Marsh (1977) from the geometry of the molecules. For larger molecules, the value of α depends on the range of densities of the data used to evaluate the constants of Eq. (6.26). For example, the constants of

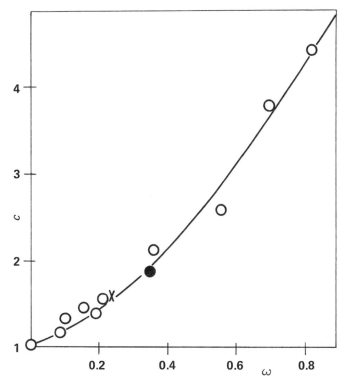

FIG. 6.4. For small and intermediate-sized molecules, the number of external degrees of freedom $3c$ is related to the acentric factor. The points were calculated by Beret and Prausnitz for Ar and hydrocarbons up to C_{20} (open circles), CO_2(X), and water (filled circles).

naphthalene were determined from the liquid densities and the vapor pressures below 1 bar, and α appears to be smaller than Eq. (6.49) suggests.

Similarly, the constant c in Eq. (6.24), where it plays a role analogous to α in (6.26), smoothly increases with the acentric factor for hydrocarbons up to C_{20}, as shown in Fig. 6.4.

The accuracy of Eqs. (6.26), (6.28), and (6.29) was extensively tested. The saturated vapor pressures and the liquid and the vapor volumes were calculated by using the Gibbs equilibrium conditions (2.11) and (2.12). As stated by Simnick, Lin, and Chao (1979), "the deviations are generally too small to be meaningfully displayed in graphs." Kreglewski found that the equation of state is valid for liquids up to

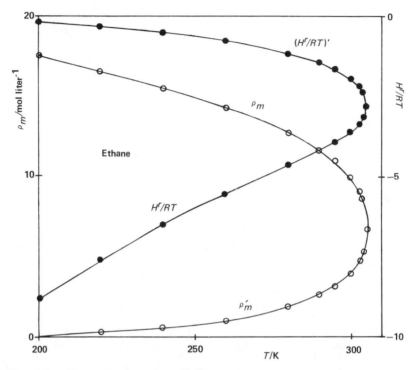

FIG. 6.5.   The molar densities ρ'_m and ρ_m of the vapor and the liquid, and the residual enthalpies of ethane. The curves represent accurately smoothed values by Goodwin et al. (1976). The points were calculated from Eqs. (6.26), (6.28), and (6.29); $T^c = 305.42$ K.

about 0.1°C below T^c. The molar densities and the residual enthalpies of ethane are shown in Fig. 6.5. The enthalpy of vaporization is

$$\Delta H^v = (H^r)' - H^r \tag{6.50}$$

The residual Helmholtz energies of the coexisting phases as functions of T are shown in Fig. 6.6. Possible errors in this function as well as in U^r/RT are assessed indirectly from the errors in ρ_m and H^r/RT. An equation of state that does not pass such a test at least in a limited temperature range may be useful for evaluation of certain properties of mixtures, such as G^E, but not H^E and the phase equilibria at high pressures.

The comparison of U^r/RT of the liquids and that of the gases at

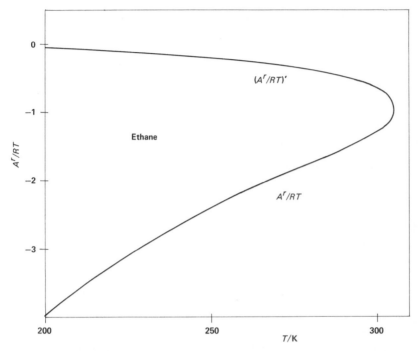

FIG. 6.6. The residual Helmholtz energies of the coexisting phases of ethane, calculated from Eq. (6.28); $T^c = 305.42$ K.

$\rho \to 0$ with the experimental data is an important test of an equation of state. Two sets of values were calculated by S. S. Chen and Kreglewski (1977). In the first one, η/k was assumed to be zero, and certain average $\bar{u} = \bar{u}^o$ values were obtained from *PVT* data at high densities. The resulting values of U^r/RT at $T/T^c = 0.6$ are given in the first column of Table 6.5, and the $\beta(T)$ curves are shown in Fig. 6.7 (dashed lines). All the values are less negative than the observed values. The second set of values was calculated by using the constants from Table 6.4, where $\eta > 0$. They agree well with the observed values of U^r/RT for the liquids. The agreement of the *predicted* $\beta(T)$ with the experimental values is excellent.[†]

[†]The properties calculated after fitting one or more constants to the experimental values of this property are often called *predicted*, whereas in fact these are only *smoothed* values. The adjective "predicted" is used by the author when the property is calculated without the knowledge of any experimental data.

TABLE 6.5.   Residual internal energies U^r/RT of liquids at $T/T^c = 0.6$ calculated from Eq. (6.29).

	Calculated[a]	Calculated[b]	Observed
Argon	−7.56	−7.56	−7.56
Nitrogen	—	−8.04	−8.05
Methane	−7.73	−7.71	−7.72
Ethane	−7.88	−9.06	−8.86
Propane	−7.94	−9.67	−9.53
n-Butane	−7.99	−10.2	−10.0
Neopentane	−8.36	−10.7	−10.1
n-Heptane	—	−12.0	−12.0
Hydrogen sulfide	—	−8.82	−8.72
Ethylene	—	−8.72	−8.59
Benzene	—	−10.4	−9.90
Carbon dioxide	—	−12.5	−11.1
Sulfur dioxide	—	−11.9	−10.7

(a)Calculated by using average values of V^{oo}, α, $\bar{u}/k$ obtained for $\eta/k = 0$.
(b)Calculated by using the constants from Table 6.4 ($\eta/k > 0$).
Note that the values calculated for C_6H_6, CO_2, and SO_2 are too negative. These deviations would be eliminated if the term $\Delta/(kT)^2$ (quadrupolar contribution) were included in $\bar{u}/k$.

These results show clearly that the temperature dependence of $\bar{u}/k$ of nonspherical or globular molecules cannot be neglected. This is perhaps the most important result of the applications of Eq. (6.26), because earlier proofs of the theory of noncentral forces for London energies are not sufficient (except the molecular dynamics results presented in Table 3.1).

The value of U^r/RT for n-heptane is correct because the constants were obtained mostly from the data for the liquid state. However, as shown in Chap. 4 and in Fig. 4.3, the calculated $\beta(T)$ curves are much less negative than the observed values. These deviations begin with n-hexane or benzene.

The second virial coefficients predicted for CO_2 and SO_2 are shown in Fig. 6.8. They tend to be correct at high temperatures. At low temperatures, even the deviations noted for CO_2 may become large. Doubtless, for strongly polar or quadrupolar fluids the Helmholtz energy given by (6.28) should be augmented by the Gubbins-Twu equations:

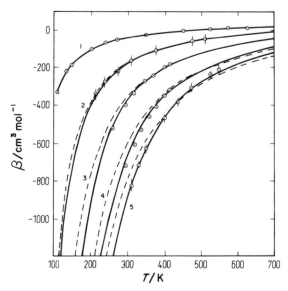

FIG. 6.7.   The second virial coefficients of (1) methane, (2) ethane, (3) propane, (4) *n*-butane, and (5) neopentane predicted by using Eq. (6.36) and the constants obtained from the data at high densities: dashed lines for $\eta = 0$ (average values of V^{oo}, α, $\bar{u}/k$), full lines for $\eta > 0$ (constants from Table 6.4). The experimental points were determined (see Dymond and Smith, 1969) by (1) Byrne et al. ($T < 273$ K) and Douslin et al. ($T > 273$ K); (2) Hoover et al. ($T < 273$ K) and Hamann and McManamey ($T > 273$ K); (3) and (4) McGlashan and Potter; (5) Dymond and Smith ($T < 400$ K) and Beattie et al. ($T > 400$ K). Reproduced with permission from *Ber. Bunsenges. Phys. Chem.* 81:1048.

$$A^r = A^r(6.28) + A^{(2)} + A^{(3)} \tag{6.51}$$

Eqs. (6.26) and (6.28) confirm the classical principle of corresponding states for small molecules (for example, $T_2^c/T_1^c = \bar{u}_{22}/\bar{u}_{11}$; see Table 6.4), but they are more general because they empirically account for the distribution functions, so (6.28) is the proper function $A^L(T^*, \rho^*)$ in the general relation (4.47). At all densities except $\rho = 0$ and except for the hydrogen-bonded molecules, Eq. (6.28) suffices for our purposes, particularly for mixtures whose properties are insensitive to the differences in the dependence of the three functions in (6.51) on ρ^*.

The next important results of applying Eq. (6.28) are the thermo-

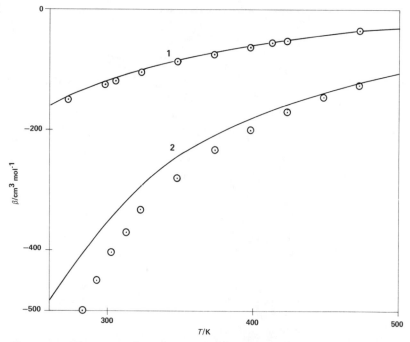

FIG. 6.8.   The second virial coefficients of (1) carbon dioxide and (2) sulfur dioxide, predicted by using the constants given in Table 6.4. The experimental points were determined by (1) Michels and Michels and (2) Kang et al. See Dymond and Smith (1969).

dynamic functions at the gas-liquid critical point. The calculated values are given in Table 6.6. They agree with the best data of Angus and Armstrong (1971, 1976), which are known for small molecules. At the critical point, U^r/RT appears to increase with the complexity of the molecules. Hence, the hypothesis of van Dranen (1952), according to which it should be constant, is not confirmed (the part due to attraction, U^{ra}/RT, is not constant either).

The residual Helmholtz energy appears to be nearly constant at the critical point. The average value is

$$A^r/RT \approx -0.99 \pm 0.02 \ (\pm 2\%); \qquad (V = V^c; T = T^c) \quad (6.52)$$

Since A^r strongly depends on ρ, the scattering of the values may be partly due to the uncertainty of ρ^c. The quantity $(Z^c - \ln Z^c)$ varies very slowly so that G^r/RT is also nearly constant:

TABLE 6.6.   Values of the residual functions at the critical point (Eqs. 6.26, 6.28, and 6.29).

	Z^c	A^r/RT^c	U^r/RT^c	G^r/RT^c
Argon	0.291	−0.944	−1.90	−0.418
Nitrogen	0.292	−0.938	−1.93	−0.416
Methane	0.285	−0.952	−1.92	−0.412
Ethane	0.280	−0.969	−2.08	−0.416
Propane	0.277	−0.972	−2.14	−0.413
n-Butane	0.274	−0.981	−2.21	−0.412
n-Pentane	0.251	−1.02	−2.34	−0.390
Neopentane	0.261	−1.006	−2.25	−0.402
n-Hexane	0.248	−1.02	−2.39	−0.381
n-Heptane	0.243	−1.03	−2.45	−0.367
n-Octane	0.235	−1.03	−2.50	−0.348
n-Decane	0.226	−1.03	−2.60	−0.316
Cyclohexane	0.265	−0.996	−2.25	−0.401
Ethylene	0.280	−0.971	−2.05	−0.420
Benzene	0.271	−0.980	−2.21	−0.404
m-Xylene	0.235	−1.04	−2.46	−0.359
Carbon dioxide	0.269	−0.978	−2.36	−0.397
Sulfur dioxide	0.269	−1.00	−2.41	−0.419
Hydrogen sulfide	0.281	−0.969	−2.04	−0.419

$$G^r/RT \approx -0.40 \pm 0.02\ (\pm 5\%); \quad (V = V^c;\ T = T^c) \quad (6.53)$$

The values of G^r/RT vary with increasing length of *n*-alkanes much more than do those of A^r/RT.

The calculated critical compressibility factor Z^c is of course correct because the conformity with the critical point was enforced. The factor Z^c is not constant, as the principle of corresponding states for small molecules (Eq. 4.1) suggests. Hougen, Watson, and Ragatz (1959) established the empirical relation

$$Z^c = 0.293/(1 + 0.375\omega) \tag{6.54}$$

for nonpolar fluids.

The constants of the equations of state (6.24) and (6.26) are known for only a few substances. Fortunately, the excess functions of mixtures can be evaluated by using estimated values of V^{oo}, $\bar{u}^o/k$, η/k, and c or α of the pure components. The constants c or α can be estimated from the values of ω. According to Beret and Prausnitz, V^o and $\bar{u}/k$ (both assumed to be independent of T) are linear functions of the van der Waals volumes and the energy of vaporization, respectively, as defined by Bondi (1968) and calculated by a group contribution method. Also, the ratio of the liquid molar volumes $V^\bullet$, at $T/T^c = 0.6$, to V^{oo} is approximately constant:

$$V^\bullet / V^{oo} \approx 1.70 \qquad (6.55)$$

The values of $V^\bullet$ are obtained by interpolation from density data (see the Appendix). The critical volume V^c is less suitable for the estimation of V^{oo}, and it is known only for the more volatile liquids.

Any one of the characteristic constants for a dense fluid, when the others are known, may be estimated by *the general method*, relying on the accuracy of Eqs. (6.26) and (6.28). If α, V^{oo}, and η/k have already been estimated, $\bar{u}^o/k$ is determined by varying its value until the liquid volume, calculated by using the Gibbs conditions (2.11) and (2.12), agrees with the observed value at a certain T—say, $T/T^c = 0.6$.

Calado and his coworkers (1978) used this method for estimating $u^L(r)$ in $u(r)$ of pure polar fluids. If μ and Q are known, then $\bar{u}^o/k$ and σ—determining $u^L(r)$, $J_{ij}^{(n)}$, and K_{ijk} in A^L, $A^{(2)}$, and $A^{(3)}$, respectively, in (4.32) to (4.36)—are found by an iteration. In this way the values of $\bar{u}^o/k$ and σ of HCl and HBr were obtained by fitting the vapor pressures and the liquid densities over a range of temperatures.

Recently, G. Iglesias-Silva and K. R. Hall (Texas A&M University, private communication, 1983) examined the deviations from Eqs. (6.26) and (6.29), using experimental data for argon up to about 10,000 bar. They have found that it is more accurate to increase the repulsion step, shown in Fig. 3.5, to $9\bar{u}^o/k$. The value of $s_1/\sigma = C_1$ is then constant over a greater range of P and T and equals 0.18. However, the most important result is that if the effective collision diameter, or V^o, is assumed to depend on ρ, the number of the universal constants D_{nm} can be decreased to about twelve. In spite of the reduction of the number of terms in the power series, the equation of state is more accurate than (6.26). The relation for A_m^r/RT, U_m^r/RT, and C_{vm}^r/R remain the same (6.28, 6.29, and 6.31), but ξ in the repulsion terms and ρ^* in the attraction terms are replaced by

$$y = V^\sigma \rho/(1 + c\, V^\sigma \rho) \tag{6.56}$$

where $V^\sigma = V^{o\sigma}[1 - C_1\exp(-9\bar{u}^o/kT)]^3 = 1/6\,\pi N_o(\sigma(T))^3$ is the volume of N_o molecules at $\rho = 0$ and $c \approx 0.45$. The dependence of y on ξ has a theoretical foundation, but the coefficient 0.45 is empirical.

According to Beret and Prausnitz, Eq. (6.24) is valid also for polymers, but the evidence is insufficient. Earlier, Flory, Orwoll, and Vrij (1964) derived an equation of state for polymers valid only for liquids composed of chain or globular molecules and below the normal boiling point. Recently, Lacombe and Sanchez (1976) derived an equally simple but more general equation:

$$Z_m = 1 - r[1 + \ln(1 - \rho_m^*)/\rho_m^* + \rho_m^*\tilde{T}_m] \tag{6.57}$$

where r is the number of segments in the r-mer and $\tilde{T}_m = kT/\varepsilon_m$, where ε_m is the interaction energy between the segments. Like all other equations in this chapter, it is formally identical for mixtures and for pure fluids. In $\tilde{T}_m$, ε_m/k is expressed by the square-well approximation (5.25), but the g_i fractions are defined as in the Flory-Huggins formula, $\varphi_i = \phi_i = r_iN_i/\sum_{h=1}^{m} r_hN_h$. In the limit $\rho \to 0$, Eq. (6.57) tends to that for the perfect gas. Hence, $\beta(T)$ cannot be calculated from this equation, although it yields good values of the vapor densities. The calculated liquid densities are too small; therefore, the values of A^r and U^r are less negative than the observed values. However, the vapor pressure curves and the excess functions of mixtures are satisfactory. The equation fails at high pressures, particularly in a large region around the gas-liquid critical point. Except in this region, the Lacombe-Sanchez equation may be useful because of its simplicity, specifically for polymer solutions.

The equations of state will be applied to mixtures in later chapters, but we may answer now why the older theories of mixtures failed in most cases. One of the reasons, already discussed in Chap. 4, was improper averaging of the pair interactions, with an excessive weight attributed to the sizes of the components. This faulty averaging was not common to all theories. The main reason for the failure was the neglect of the repulsion terms in the thermodynamic functions. The values of all the relevant terms, calculated from (6.28) for saturated liquid argon at $T/T^c = 0.6$ (that is, below the normal boiling point) and at the critical point, are as follows:

T/T^c	Z^h	Z^a	Z	A^{rh}/RT	A^{ra}/RT	A^r/RT
0.6	7.49	−7.49	0.00	3.31	−7.60	−4.29
1.0	1.983	−1.692	0.291	0.791	−1.735	−0.944

T/T^c	U^{rh}/RT	U^{ra}/RT	U^r/RT
0.6	0.08	−7.64	−7.56
1.0	0.05	−1.95	−1.90

Because $(\partial\xi/\partial T)_v$ is small, U^{rh}/RT is small, and it is entirely negligible in the excess functions for mixtures. Consequently, the theories made it possible to estimate the excess enthalpies H^E. Such correlations are reviewed by Kehiaian (1972). The *lattice theories* are most clearly outlined by Hill (1960) and by Prigogine and his coworkers (1957). These theories operate with the so-called *interchange energies* $w = \bar{u}_{11} + \bar{u}_{22} - 2\bar{u}_{12}$, which are independent of the distances r and are thus the minimum values for the square-well or the Kac-Uhlenbeck-Hemmer potential. In the Bragg-Williams model, w is independent of T, and the mixture is a random one. All sites on the lattice are occupied, and therefore the model corresponds to a completely condensed fluid with $P = 0$. In this model, at a constant T,

$$U^E/RT = H^E/RT = x_1 x_2\, zw/2RT \tag{6.58}$$

where z is the number of nearest neighbors (the coordination number) assumed to be independent of T; thus,

$$G^E = A^E = U^E \quad \text{because} \quad S^E = -(\partial G^E/\partial T)_P = 0 \tag{6.59}$$

Guggenheim (1952) introduced the *quasi-chemical model*, which takes into account the possibility that differences in $\bar{u}_{ii}$ and $\bar{u}_{jj}$ will tend to favor the distributions that lower the lattice energy. In this model,

$$\frac{A^E}{RT} = \frac{1}{2}\, x_1 x_2\, \frac{zw}{kT}\left(1 - x_1 x_2\, \frac{w}{2kT} + \ldots\right) \tag{6.60}$$

and S^E is always negative. The correction due to the deviations from randomness of mixing appears to be negative but small, less than 5 percent of total A^E. Because of the presence of noncentral energy

$(\eta_{ij}/k > 0)$, the excess entropies are usually positive and become negative only in special cases. This does not diminish the value of Eq. (6.60). In this elegant treatment, the deviations due to nonrandom mixing are evaluated exactly for a rigid lattice and remain valid for real fluids. However, (6.60) cannot be applied to real fluids as it stands. Eqs. (6.60) and (6.58) have been widely used to estimate the contributions of London energies in systems with specific interactions. The value of z was usually kept constant; however, it is a function of ρ^*. Eq. (6.60) is the mixed, $1 + 2$, term in A_m^{ra}/RT. If z is proportional to ρ^*, A_m^{ra}/RT corresponds to that from (6.18). For the square-well approximation, we obtain

$$\frac{A_m^r}{RT} = \frac{A_m^{rh}}{RT} - \frac{c\rho_m^*}{kT}\left[x_1\bar{u}_{11} + x_2\bar{u}_{22} - x_1x_2w\left(1 + \frac{x_1x_2w}{2kT}\right)\right] \quad (6.61)$$

where c is a dimensionless constant that may be fitted to the properties of argon. The values of V^{oo} and $\bar{u}/k$ of the pure components of a given system are then obtained by the general method outlined following Eq. (6.55) (eventually, putting $\alpha = 1$ for all substances). They will differ from those given in Table 6.4. The range of validity of (6.61) is restricted to high densities, but the excess functions will be more accurate than those calculated from (6.60), particularly when $\bar{u}_{11}$ and $\bar{u}_{22}$ are very different (if $\bar{u}_{11} \ll \bar{u}_{22}$, then nearly always $\rho_1^* \ll \rho_2^*$ for the saturated liquids at constant T and $k_{12} \neq 1$ in the rule 5.13).

The lattice model led Becker and Kiefer (1973) to a relation for the upper critical solution temperature:

$$G_c^E/RT^{ucs} \approx 0.55 \quad (6.62)$$

where G_c^E is the value of the excess Gibbs energy at the critical point, $T = T^{ucs}$, and $x_i = x_i^c$. When G^E is known, the relation allows one to guess whether the mixture is homogeneous or whether two liquid phases will appear.

Deiters (1983) extended the quasi-chemical model to mixtures of spheres of different sizes and obtained the relation

$$\bar{u}_m = x_1\bar{u}_{11} + x_2\bar{u}_{22} + x_1s_1q_2\,\Delta\varepsilon Q \quad (6.63)$$

where $\Delta\varepsilon = (2\bar{u}_{12}/s_{12}) - (\bar{u}_{11}/s_1) - (u_{22}/s_2)$, the function Q is the nonrandomness correction, and s_i and q_i are the contact numbers and the contact fractions, respectively (see Eq. 5.20). For $Q = 1$, Eq. (6.63) is identical in form with Eq. (5.24).

The most accurate known equations of state contain the cumbersome power series (6.23). There are ways to improve this state of affairs. But almost certainly the starting point for any theory of fluid mixtures must be the van der Waals theory in its modern form.

CHAPTER 7

Phase Equilibria at High Pressures:
The Critical State

A. The Critical States of Pure Fluids and Mixtures

The critical phenomena of liquids have fascinated many physical chemists for more than a century. They were first observed by Cagniard de la Tour (1822), but research under more controlled conditions was stimulated much later by the works of Andrews and van der Waals in the 1870's.

When a proper amount of a liquid (such that the average density of the vapor and the liquid approximately equals the critical density) is confined in a glass tube sealed at both ends and heated to the critical temperature T^c, the meniscus vanishes, and in its place a narrow band appears. If the tube is kept at rest and a constant $T = T^c$, the band may persist for many hours. When the tube is rotated or its contents stirred with a steel ball, the band vanishes, and the whole fluid exhibits a brownish opalescence well visible in a yellow light. Its intensity rapidly vanishes when the temperature departs even slightly from T^c. According to the theory of Smoluchowski (1908) and Einstein (1910), the point of maximum opalescence corresponds to the critical point of a liquid. It was experimentally confirmed by Mason, Naldrett, and Maass (1940) and by Naldrett and Maass (1940) that the points are exactly identical ($\pm$ 0.001°C).

Under the conditions described above, critical phenomena in mixtures are very similar to those observed in pure fluids. However, if one end of the tube is sealed by means of mercury so that the volume of the fluid above it can be varied, a characteristic difference between the behavior of pure fluids and that of mixtures becomes evident, particularly when the critical temperatures of the components are very different. Instead of the single vapor pressure-temperature, $P(T)$, curve of a pure fluid shown in Fig. 7.1(A), every mixture of constant composition (except an azeotropic mixture) will have two lines, the bubble-point and the dew-point lines, that meet at the *critical point*, as shown

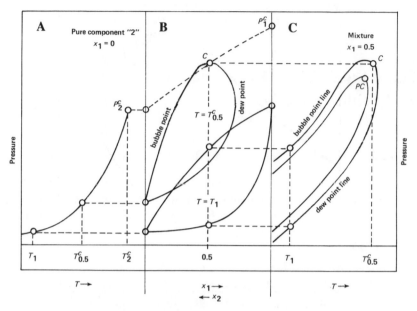

FIG. 7.1.   The differences between the behavior of pure components and that of a binary system.

in Fig. 7.1(C) and the enlarged detail in Fig. 7.2. The point PC is the pseudocritical point discussed in Section C of this chapter. Fig. 7.1(B) shows the boiling (bubble) and condensation (dew) pressures at two temperatures—the first $T = T_1$, lower than any T^c in this system, and the second $T = T^c_m$ ($x = 0.5$)—as functions of x_1 briefly called the equilibrium isotherms.

When the mixture is compressed *isothermally* at $T > T^c_m$ from state A to state D (Fig. 7.2), the amount of the liquid phase initially increases up to state B. When compressed further, the liquid level remains constant, and a narrow band of opalescence appears. Stirring causes the opalescence to disperse over the whole system, and the amount of the liquid phase *decreases*. No heat effect is noticed in this process, and an amount of the liquid phase is apparently *dispersed* in the vapor phase. At point D all the liquid phase vanishes. The two phases in the shaded area were called by Swietoslawski (1945) the *liquidosol* and the *gaseosol*. These names appropriately reflect the state of mutual dispersion of the phases.

A similar but reverse phenomenon may be observed when the

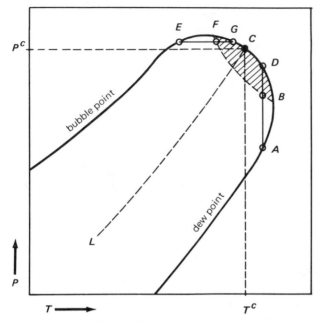

FIG. 7.2.   Enlarged portion of Fig. 7.1(C), showing retrograde condensation in a system at constant composition. Line L is the vapor pressure curve of a pure substance with the same critical point C.

state of the system approaches the vicinity of the critical point by *iso-baric* heating at densities somewhat larger than the critical density— say, along the line E-F-G. In state E, the tube is completely filled with the liquid phase (bubble point). When the mixture is heated, the me-niscus level drops to a minimum at point F. With further heating *and stirring*, the meniscus rises until the next bubble point G is reached, and the tube is completely filled with the liquidosol.

The phenomena along lines B-D and F-G were first observed by Kuenen (1906) in 1892; he called them *retrograde condensation* of the first kind and the second kind, respectively.[†] Kuenen also gave the first explanation based on classical thermodynamics. It is based on the shape of the $U(S, V)$ surface and is difficult to "translate" to observ-able properties. The clearest explanation based on the $A(T, V)$ surface is due to Rowlinson (1969). Briefly, if the Helmholtz energy is a con-

---

[†]From the physical point of view, a better term might be the *critical dispersion* of the phases.

tinuous function of T, V, and x_i at and near the critical point, then the pressures $[= -(\partial A/\partial V)_T]$ along the dew- and bubble-point curves must form a rounded end of the loop, as shown in Fig. 7.2. The derivatives $(\partial P/\partial x_i)_{T,\sigma}$ along the two saturation curves (subscript σ) must become identical at the critical point as well as $(\partial T/\partial x_i)_{P,\sigma}$ along the $T(x_i)$ loop. One of the important consequences is that the dew- and bubble-point curves on the $P(T)$ diagram at constant x_i cannot cross each other at the critical point but have to meet, forming a rounded end of the loop. Coexisting phases, therefore, must exist at $T > T_m^c$ or $P > P_m^c$, although it may be a very small range of T or P. The critical point C is sometimes shifted to the vicinity of point F, and the second kind of retrograde condensation nearly vanishes, whereas the region of the first kind becomes very large.

An exception is an azeotropic mixture. Here, the dew- and bubble-point pressures are identical; the $P(T)$ loop along the azeotropic composition reduces to a single curve like that of a pure substance, and retrograde condensation does not appear.

Differences between pure fluids and mixtures are also evident on the $\rho(T)$ graphs at constant x_i. They are shown in Fig. 7.3. While the curves for the pure components are symmetrical, those of the mixtures exhibit a hump on the dew-point curve ($\rho_m < \rho_m^c$). The shaded area above T_m^c at $x_1 = 0.9699$ obviously corresponds to the region of retrograde condensation of the first kind. The dashed line is the line of the critical points, also called the *critical locus curve*. If the densities were plotted against the liquid-vapor equilibrium pressures, $\rho(P)$ at constant x_i, the curves for mixtures would exhibit very small humps, corresponding to retrograde condensation of the second kind.

The behavior in the gravitational field is also different. The compressibility $(\partial \rho/\partial P)_{T^c, \rho^c}$ is infinitely large for a pure fluid. Consequently, a small pressure difference in the system may create a large density gradient. The weight of the fluid in the upper part of the vertically held experimental tube suffices to compress the lower layers so that their density increases. The effect was accurately measured in the pioneering works of Naldrett and Maass (1940), Weinberger and Schneider (1952), and Habgood and Schneider (1954). Later studies are reviewed by Sengers and Levelt-Sengers (1968) and by Hohenberg and Barmatz (1972). Fig. 7.4 shows the coexistence curves of highly purified xenon near the critical temperature. Both curves were obtained for a carefully stirred sample, keeping the temperatures con-

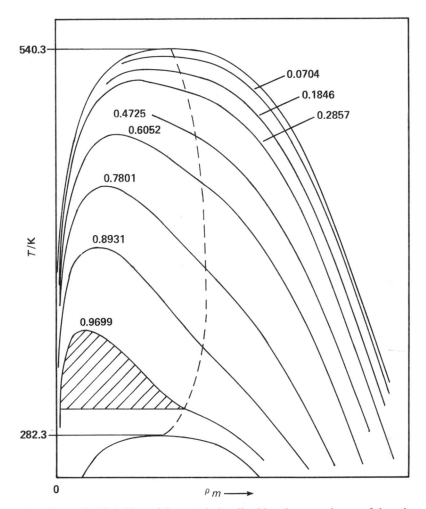

FIG. 7.3.   The densities of the coexisting liquid and vapor phases of the eth-ylene (1) + *n*-heptane (2) system determined by Kay (1948). The dashed line is the critical locus curve. The numbers indicate the values of x_1. The experi-mental points are deleted.

stant within 0.001°C. The curve obtained when the tube (length 19.5 cm) was kept vertical has a flat top, whereas that in which the gravity effect was largely eliminated by using a short bomb (1.2 cm) has a rounded top.

If the system is a mixture, $(\partial\rho/\partial P)_{T^c,\rho^c}$ has a finite value, and the

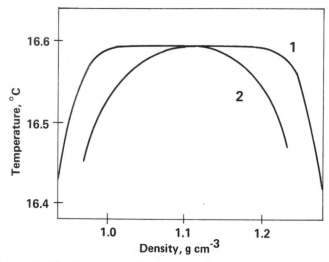

FIG. 7.4.   The liquid-vapor coexistence curves of xenon in the vicinity of T^c (16.583°C) determined by Weinberger and Schneider (1952): curve 1 in long bomb; curve 2 in short bomb (or the bomb in horizontal position). The experimental points are deleted.

effect is much smaller; however, a concentration gradient may occur. These are considered later.

The work of Schneider and coworkers confirmed the classical view of the critical state as the point at which the densities of the coexisting phases become equal and as the highest temperature at which $(\partial P/\partial \rho)_T$ is zero for a pure fluid. The surface tension tends to zero at this point, but this property has a clear meaning only when the interface is well defined. A few degrees below T^c the surface of the liquid appears to be flat, but very near to T^c it is completely diffuse. The interface at T^c is that between a vapor and clusters ranging in size from 10^2 to 10^5 molecules. A physical theory of fluid interfaces is due to Sarkies, Richmond, and Ninham (1972), Weeks (1977), and others quoted in these papers. The gas-liquid and the gas-solid interfaces have been studied by molecular dynamics and Monte-Carlo simulations by Freeman and McDonald (1973), Saville (1977), and Chapela et al. (1977). The thickness of the interface appears to be less than 2σ near the triple point and increases rapidly as the critical point is approached. The same studies for an equimolar mixture show that the more volatile component is adsorbed on the interface, forming a layer one σ thick in the gas phase.

The thermodynamic conditions of a critical point were derived long ago by Gibbs and by van der Waals and his coworkers, but the procedure is most clearly outlined by Rowlinson (1969) and Münster (1970). If the Helmholtz energy at the critical point is a continuous function of V and x_i, then the conditions for an m-component system are

$$D_A = \begin{vmatrix} \dfrac{\partial^2 A_m}{\partial V_m^2} & \dfrac{\partial^2 A_m}{\partial V_m \partial x_1} & \cdots & \dfrac{\partial^2 A_m}{\partial V_m \partial x_{m-1}} \\ \dfrac{\partial^2 A_m}{\partial x_1 \partial V_m} & - & & \vdots \\ \vdots & & & \vdots \\ \dfrac{\partial^2 A_m}{\partial x_{m-1} \partial V_m} & \cdots & \cdots & \dfrac{\partial^2 A_m}{\partial x_{m-1}^2} \end{vmatrix} = 0 \qquad (7.1)$$

The determinant D_A', formed from (7.1) by replacing the elements of an arbitrarily chosen row by

$$\frac{\partial D_A}{\partial V_m}, \frac{\partial D_A}{\partial x_1}, \quad \cdots \quad , \frac{\partial D_A}{\partial x_{m-1}}$$

is also zero:

$$D_A' = 0 \qquad (7.2)$$

All the partial derivatives are formed at constant T. For a binary system, the conditions conveniently transformed by Teja and Rowlinson (1973) and Teja (1975) are

$$K\left(\frac{\partial^2 A_m}{\partial x_1^2}\right)_{T,V} - \left(\frac{\partial^2 A_m}{\partial x_1 \partial V_m}\right)_T = 0; \qquad (7.3)$$

$$K^3\left(\frac{\partial^3 A_m}{\partial x_1^3}\right)_{T,V} - 3K^2\left(\frac{\partial^3 A_m}{\partial x_1^2 \partial V_m}\right)_T$$

$$+ 3K\left(\frac{\partial^3 A_m}{\partial x_1 \partial V_m^2}\right)_T - \left(\frac{\partial^3 A_m}{\partial V_m^3}\right)_{T,x} = 0 \qquad (7.4)$$

where

$$K = \left(\frac{\partial^2 A_m}{\partial V_m^2}\right)_{T,\,x}\left(\frac{\partial^2 A_m}{\partial x_1 \partial V_m}\right)_T^{-1}$$

Using the critical conditions in this form avoids computational diffi-
culties that otherwise occur for the compositions for which $\partial^2 A_m/\partial V_m^2$
may become very small. The conditions are equivalent to the familiar
conditions derived in textbooks on thermodynamics for a binary system:

$$\left(\frac{\partial^2 G_m}{\partial x_1^2}\right)_{T,\,P} = \left(\frac{\partial^3 G_m}{\partial x_1^3}\right)_{T,\,P} = 0 \tag{7.5}$$

or the J. W. Gibbs (1876) conditions:

$$\left(\frac{\partial \mu_1}{\partial x_1}\right)_{T,\,P} = \left(\frac{\partial^2 \mu_1}{\partial x_1^2}\right)_{T,\,P} = 0 \tag{7.6}$$

or those derived from the conditions for partial molar quantities (see,
for example, R. L. Scott, 1978):

$$\left[\frac{\partial(\mu_2 - \mu_1)}{\partial x_1}\right]_{T,\,P} = \left[\frac{\partial^2(\mu_2 - \mu_1)}{\partial x_1^2}\right]_{T,\,P} = 0 \tag{7.7}$$

For a pure fluid, the derivatives $\partial^n/\partial x_i^n$ ($n = 1, 2, 3$) are zero, and
(7.1) and (7.2) simplify to the following conditions:

$$\left(\frac{\partial^2 A_m}{\partial V_m^2}\right)_T = -\left(\frac{\partial P}{\partial V_m}\right)_T = 0 \tag{7.8}$$

and

$$\left(\frac{\partial^3 A_m}{\partial V_m^3}\right)_T = -\left(\frac{\partial^2 P}{\partial V_m^2}\right)_T = 0 \tag{7.9}$$

When the conditions (7.3) and (7.4) are expressed by an accurate
equation of state, the relations obtained are usually complicated, and
the iterations may not converge in some cases to the equilibrium values
of P^c, ρ_m^c, x_1^c, x_2^c, . . . for a chosen $T = T^c$. Fortunately, the critical

locus of a mixture can be calculated without the use of those conditions. As shown by Rowlinson (1969), the derivatives $(\partial P / \partial x)_{T, \sigma}$ along the liquid-and-vapor or liquid-and-liquid saturation curves (subscript σ) are

$$
\left(\frac{\partial x_i}{\partial P} \right)_{T, \sigma} = - \left(\frac{\partial x_i}{\partial P} \right)'_{T, \sigma}
$$

$$
= -6 \left(\frac{\partial^2 V}{\partial x_i^2} \right)^c_{P, T} \left[\Delta x \left(\frac{\partial^4 G}{\partial x_i^4} \right)^c_{P, T} \right]^{-1} \quad (7.10)
$$

where $\Delta x = (x_i - x_i')$ and vanishes at the critical point. Analogous relations hold for $(\partial T / \partial x_i)_{P, \sigma}$ and $(\partial T / \partial x_i)'_{P, \sigma}$. Hence, the critical point is always at an extreme value of $P(x)$ at constant T, or of $T(x)$ at constant P. The bubble points (ρ_m, x_i) and the dew points (ρ_m', x_i') at constant T are obtained by varying P in small increments until the calculated densities and compositions of the phases become nearly identical. It suffices to estimate the gas-liquid P^c within $\pm$ 0.1 bar and the gas-gas P^c within $\pm$ 10 bar and x_i^c within $\pm$ 0.01 because the uncertainty of the experimental or the calculated values is usually greater. The value of ρ_m^c is then readily calculated.

Certain mixtures may separate into two or more liquid phases. The temperatures at which the compositions and the densities of the phases become identical and the liquids again become completely miscible are called the *upper* (T^{ucs}) and *lower* (T^{lcs}) *critical solution temperature*. They are marked in Fig. 7.5 by points C_2 and C_1, respectively, and the dashed curve shows the miscibility gap at saturation pressure. The critical temperatures vary with the pressure, and the curve connecting the UCS and LCS points in the (P, T, x_i) space is called the *liquid-liquid critical locus curve*. In this case, the miscibility gaps form a dome with a maximum point M. In rare cases, the limited miscibility may appear again at very high pressures with other critical points N_i (fluid-fluid critical locus curve). Most systems with limited miscibilities exhibit only UCS points. With decreasing temperature, the miscibility gap widens until the system reaches the freezing temperature. The singular point of the solid-liquid-liquid-vapor equilibrium is called the *quadruple point*. With increasing pressure, either a dome is formed or the miscibility gap widens. The possibility of the appearance of gaps and their shape depend in a very sensitive manner

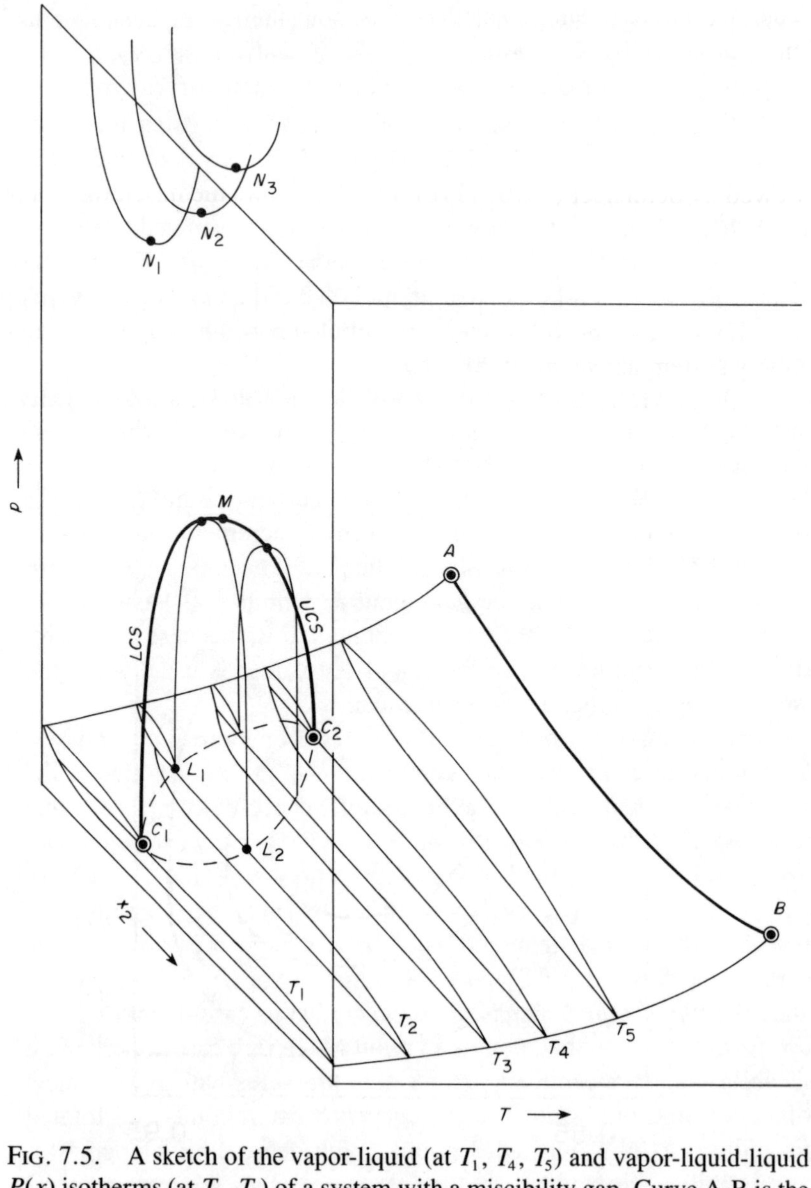

FIG. 7.5.   A sketch of the vapor-liquid (at T_1, T_4, T_5) and vapor-liquid-liquid $P(x)$ isotherms (at T_2, T_3) of a system with a miscibility gap. Curve A-B is the gas-liquid critical locus curve.

upon the kind and energy of the interactions between the components. The miscibility gaps calculated for a model system are considered in Chap. 9.

The effects of pressure on miscibility have been extensively studied by Timmermans and by Schneider and their coworkers and are reviewed by Schneider (1970, 1978). Fig. 7.6 shows the miscibility gaps of the heavy water + 2-methyl-pyridine system determined by Garland and Nishigaki (1976). The differences between T^{ucs} and T^{lcs} decrease with increasing pressure, and at about 200 bar the two become identical. However, as shown by Schneider, limited miscibility appears again in this system above about 2000 bar.

As shown by Davenport and Rowlinson (1963), a lower critical solution point (LCSP) may appear at high pressures in the methane

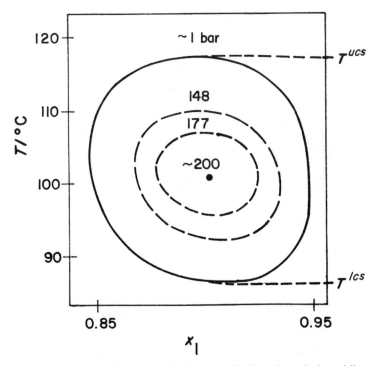

FIG. 7.6. The liquid-liquid equilibria in the D_2O + 2-methyl-pyridine system, determined by Garland and Nishigaki (1976). The miscibility gap appears on the water-rich side of the diagram. The experimental points are deleted.

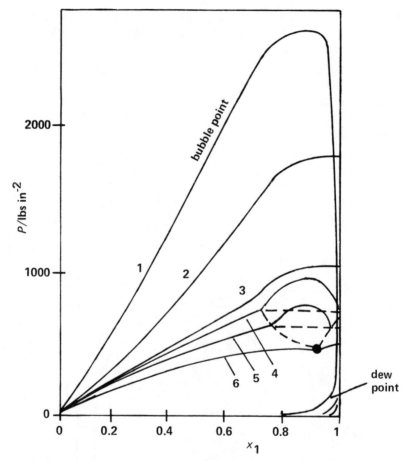

FIG. 7.7.   Six vapor-liquid isotherms of the methane (1) + *n*-hexane (2) system, taken from the data of Lin et al. (1977): (1) 273.16 K; (2) 223.15 K; (3) 198.05 K; (4) 195.91 K; (5) 190.50 K; (6) 182.46 K. The filled circle indicates the LCSP. The experimental points are deleted.

+ hexane (*n*- or isomers) systems. In binary systems with methane, olefins are less miscible than paraffins and di-olefins, and acethylenes are less miscible than olefins. Aromatics are very insoluble. Fig. 7.7 shows the liquid-vapor and liquid-liquid-vapor equilibrium isotherms of the methane + *n*-hexane system determined by Lin et al. (1977). The vapor pressures of pure hexane are very low at these temperatures, so all the lines seem to converge to one point at $x_1 = 0$. Only

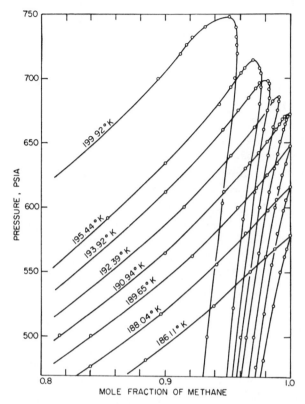

FIG. 7.8.   Enlarged part of the $P(x)$ diagram of the methane (1) + ethane (2) system. Reproduced with permission from *J. Chem. Eng. Data* 17:9.

two of the isotherms, numbers 5 and 6, reach $x_1 = 1$, because they are below the critical point of CH_4 ($T^c = 190.58$ K; $P^c = 46.04$ bar $= 667.8$ lbs $\cdot$ in^{-2}).[†] The other ones form continuous loops of bubble- and dew points, although they pass very near to $x_1 = 1$. The dashed curves show the miscibility gap, and the horizontal dashed lines are the so-called *tie lines* connecting two liquid phases in equilibrium at constant T and P.

Fig. 7.8 is an enlarged part of the $P(x)$ diagram at $x_1 = 0.8$ to 1 of another system, methane + ethane, determined by Wichterle and Kobayashi (1972). The isotherm at the critical temperature T_1^c of the

[†]The critical constants of pure fluids were selected by Kudchadker, Alani, and Zwolinski (1968), Mathews (1972), and Ambrose and Townsend (1975).

more volatile component has a peculiar shape at $x_1 \rightarrow 1$. As demonstrated by Wichterle and his coworkers (1971), the bubble-point curve $P(x)$ and the dew-point curve $P(x')$ have a common slope at this point; that is, $\lim (dx'/dx)_{T_1^c} = 1$, and the slope is $(dP/dx_1)_{T_1^c} = (dP/dx_1')_{T_1^c} = 0$. As shown by Rowlinson (1969) and Hala (1975), this behavior is required to meet in the limit the critical conditions of a pure fluid.

Phase behavior of the type illustrated in Fig. 7.7 is important because it probably exists in underground petroleum deposits and may have a wide application in the extraction of the liquid components of a mixture by means of compressed gases.[†] Phase equilibria in petroleum are immensely complicated by the presence of water, salt, carbon dioxide, and hydrogen sulfide. As is reviewed by Schneider (1978), hydrocarbon systems with these compounds are being extensively studied. The first works on the effects of H_2S and CO_2, in which all the variables P, V, T, and x_i were accurately measured, are due to Kay and coworkers (Kay and Rambosek, Bierlein and Kay, Kay and Brice, all in 1953), Poettmann and Katz (1945), and Reamer, Sage, and Lacey (1951).

When the ratio of the critical temperatures T_2^c/T_1^c or $\bar{u}_{22}/\bar{u}_{11}$ increases, the (usually) continuous gas-liquid critical curve may break into two curves, as shown in Fig. 7.9. One of them extends from the critical point of the more volatile component A to the *upper critical end point* (UCEP) E_1. At this point two of the phases rich in component A become identical. The second critical curve passes continuously from the critical point of the pure component B, through gas-liquid (C_1, C_2) and liquid-liquid (C_3, C_4) critical points to the *lower critical end point* (LCEP) E_2. The appearance of the miscibility gap here has other causes than those of the gap far below the gas-liquid critical locus (see Fig. 7.5). The gap shown in Fig. 7.5 may appear when the mixed interactions ε_{ij} are weak compared to the harmonic mean of ε_{ii} and ε_{jj}. The gap shown in Fig. 7.9 is due to the vicinity of point E_1, where the density of the more volatile component (the "solvent") rapidly decreases. The "solute" separates in a similar manner as it would from a solution in a highly compressed gas upon decompression. The UCEP usually appears at a pressure and a temperature just above P_1^c and T_1^c of the more volatile component 1 and at

[†]Phase equilibria including a solid phase and the solubility of solids in compressed gases are beyond the scope of this book. These equilibria are reviewed by Rowlinson and Richardson (1959), Ricci (1966), Prausnitz (1965), Rowlinson (1969), and Schneider (1978).

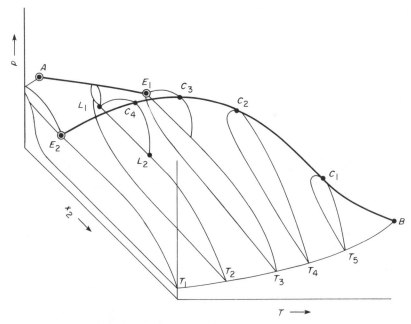

FIG. 7.9.   A sketch of the P-T-x_2 diagram of a system with two critical locus curves: A to E_1 (UCEP) and B to E_2 (LCEP).

$x_1 > 0.85$. The LCEP was first observed by Kuenen and Robson (1899) for the ethane + ethanol system, but it may also appear in systems without specific interactions, such as methane + hexane or isomeric hexanes or heptanes.

The transition from the UCEP and the LCEP to a *tricritical point*, at which three phases become simultaneously identical, was studied by Creek et al. (1977) and Specovius et al. (1981). The thermodynamics of tricritical points was developed by Griffiths (1974), Bartis (1973), Guerrero, Rowlinson, and Morrison (1976), Desrosier et al. (1977), and Fox (1978).

The thermodynamic conditions for the upper or lower critical solution points are the same as for the gas-liquid critical points given by Eqs. (7.1) to (7.10). According to Block et al. (1977) and Knobler and Scott (1978), $(\partial \mu_i / \partial x_i)_{T,P,\phi}$ and $(\partial^2 \mu_i / \partial x_i^2)_{T,P,\phi}$ in a constant gravitational field ϕ are not affected by this field.

As the ratio $\bar{u}_{22} / \bar{u}_{11}$ increases further, a gradual transition from the type of phase diagram shown in Fig. 7.9 to that shown in Fig. 7.10

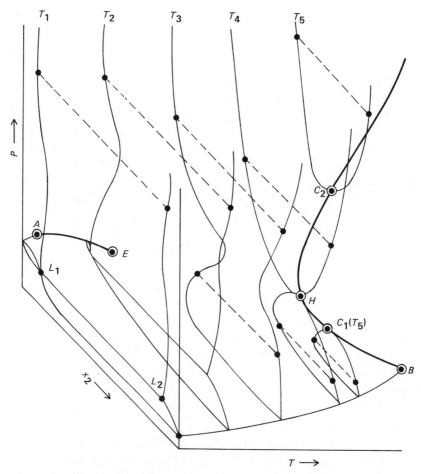

FIG. 7.10.   A sketch of the P-T-x_2 diagram of a system showing gas-gas equilibria of the second type and, at lower temperatures, a limited miscibility of the liquid phases. The dashed lines are the tie lines connecting phases in equilibrium at a constant T and an arbitrary constant P.

may occur. Point E (UCEP) moves closer to A. (Relative to the difference between T_B^c and T_A^c, the length of A-E is out of proportion; also, the L_1-L_2 equilibrium may appear at a much lower T.) The second critical locus curve may be "normal" from the critical point of component B up to a point H and then turn up to increasing temperatures. Below this point H and at temperatures between T_B^c and T_4, the liquid-vapor isotherms form the usual loops, such as T_5 with a critical point

C_1. At higher pressures two phases may again appear at the same T_5 with a critical point C_2, but the mole fraction x_2 at this point is smaller than at C_1. The states above point H were called by Tsiklis *gas-gas equilibria*. Their existence was predicted by van der Waals (1894) and also considered by Kammerlingh Onnes and Keesom (1906) on the basis of the van der Waals equation of state. The importance of these works was overlooked until Krichevski (1940), Krichevski and Tsiklis (1943), and Tsiklis (1952) observed it in the methane + ammonia and other systems. The gas-gas equilibria are reviewed by Tsiklis and Rott (1967), Rowlinson (1969), and Schneider (1978). Equilibria of the *first* and the *second* types are distinguished. Van der Waals called point H the "binary homogeneous point," but it would perhaps be less confusing to call it the *double critical point*. The gas-gas equilibria of the second type are shown in Fig. 7.10 and appear, for example, in the systems $N_2 + NH_3$ and $N_2 + SO_2$. The first type differs in that the gas-gas critical locus curve begins immediately at the critical point of the less volatile component. That is, points H and B are identical in this case. It was first observed by Tsiklis (1952) in the He + NH_3 system. The most interesting system exhibiting this type of phase diagram is the He + Xe system, studied by de Swaan Arons and Diepen (1966), because it is a system with London energies only, and the values of $\bar{u}_{ij}$ and σ_{ij} are well known (see Tables 5.1 and 5.2). About fifty systems are known that exhibit gas-gas equilibria of the first or the second type. The references were compiled by Schneider (1978) and by Hurle, Jones, and Young (1977).

There arises the question of whether the gas-gas critical P_m^c and T_m^c may increase without limits or whether they may eventually make their way back to the critical point A (Fig. 7.10). We note that no such curve was ever found by starting from A or E, where the pressures are low enough to be measured. Further, the shape of the isotherm T_3 results from two opposite effects on the Helmholtz energy of the system, given by, say, Eq. (6.28). As the pressure increases along both branches of the isotherm, the difference in ρ^* and A_m^r of the phases diminishes. If the values of $\bar{u}_{11}$, $\bar{u}_{22}$, and $\bar{u}_{12}$ are not too different, the loop may close and a gas-liquid critical point may appear. However, when $\bar{u}_{22}/\bar{u}_{11} \gg 1$ and $\bar{u}_{12}$ is relatively weak, A_m^r at the same densities ρ^* is less negative than in the former case, and a "neck" appears as on the T_3-isotherm. Upon further increase of pressure, phase separation is enhanced by rising (positive) A_m^{rh}. There is no reason to believe that the gas-gas critical locus turns around at high ρ^* toward lower temperatures; more likely, it tends to a constant temperature. This point of view will probably never be verified experimentally. As shown in the pioneering works of Bridgman, reviewed by Bradley (1963), the chemical nature of a substance may be affected profoundly by extreme pressures, and the phase equilibria measured under these conditions might be those of an altered substance.

The phenomena observed in the vicinity of gas-gas or liquid-liquid critical solution points are similar to those observed in the gas-liquid critical state. The densities of the phases are less affected by gravitational forces than those in a pure fluid. However, there are significant concentration gradients on and near the liquid-liquid critical locus curve. These were exhaustively studied by Fannin and Knobler (1974), Kwon, Kim, and Kobayashi (1977), and others; they are reviewed by R. L. Scott (1978).

According to Rice (1950) and Gopal and Rice (1955), the liquid-liquid miscibility curve exhibits a flat top at T^{ucs}, similar to that shown in Fig. 7.4. For the perfluoromethylcyclohexane + carbon tetrachloride system, and the precision of temperature measurements within $\pm 0.001°$, the flat top extends over volume fractions of CCl_4 from about 0.537 to 0.563. Kreglewski (1963) found that the mole fractions of the coexisting phases, x_1 and x_1', of the $n\text{-}C_8H_{18} + n\text{-}C_8F_{18}$ system in the range of $T^{ucs} - T$) from zero to about $2°$ closely follow the formula

$$|x_1' - x_1| = 0.187(T^c - T)^{1/2} - 0.20(T^c - T)^{1/4}$$
$$+ 0.275(T^c - T)^{1/6} \tag{7.11}$$

where $T^c = T^{ucs} = 348.61$ K. The powers $1/y$, $y = 2, 4, 6, \ldots$ have been chosen in accordance with the classical theory of critical phenomena outlined by Levelt-Sengers (1970). Suppose the uncertainty of the temperature measurements is $0.001°$, or ± 0.0005. Eq. (7.11) yields $|x_1' - x_1| = 0.057$ for $(T^c - T) = 0.001°$. That is, an apparent flat top appears, extending over this range of x_1 at a "constant" $T = T^c \pm 0.0005°$. According to recent theories, mentioned later, the liquid-vapor as well as the liquid-liquid miscibility curves have a rounded top; however, their shape is not exactly classical.

The phase diagrams near to the critical states may be complicated further by the presence of various types of azeotropic mixtures. As shown by Rowlinson (1969), the lines of azeotropic points, if present, end tangentially on the critical locus curves.

On the basis of an analysis of the phase diagrams by means of the van der Waals equation of state, R. L. Scott and van Konynenburg (1970; see also Scott, 1972) distinguished nine major types of binary systems.

(1) One gas-liquid critical line, $0 \leq x_1 \leq 1$, extending from $C_1(T_1^c, P_1^c, x_1 = 1)$ to $C_2(T_2^c, P_2^c, x_1 = 0)$.

(1a) Same as (1), with the addition of a negative azeotrope.

(2) Two critical lines: gas-liquid $0 \leq x_1 \leq 1$; liquid-liquid extending from a limiting UCSP $C^\infty(T^\infty, P = \infty, x_1^\infty)$ of a close-packed system ($\rho^* = 1$) to the upper critical end point UCEP (L-L).

(2a) Same as (2), with the addition of a positive azeotrope.

(3) Two critical lines: gas-liquid C_1 to UCEP; liquid-liquid $x_1^\infty \geq x_1 \geq 0$ extending from C^∞ to C_2.

(3a) Same as (3), except that the three-phase line lies at lower pressures than either pure component producing a "heteroazeotrope" type of diagram.

(4) Three critical lines: C_1 to UCEP (essentially G-L); liquid-liquid from LCEP to C_2; liquid-liquid from C^∞ to UCEP.

(5) Two critical lines: C_1 to UCEP (essentially G-L); liquid-liquid from LCEP to C_2.

(5a) Same as (5), with the addition of a negative azeotrope.

The various types of phase equilibria are also presented on well-conceived graphs by Wichterle (1977, 1978). The classification presented above is similar to Rowlinson's. Both are phenomenological classifications because the van der Waals equation does not produce values of A_m^r/RT suitable for quantitative comparisons. Very subtle changes in A_m^r due to an altered balance of intermolecular forces have a profound effect on the phase diagrams.

All equations of state, even the empirical ones with forty or more constants, are less accurate in the vicinity of a critical point than they are in other regions of the *PVT* diagram. For brevity, we shall call them *analytical equations*, because the behavior of fluids near the critical point is now regarded as "nonanalytical." Typical deviations of the analytical equations of state are shown schematically in Fig. 7.11. All the isotherms meet at $\rho = 0$ where $P = 0$. They are S-shaped within the liquid-vapor saturation curve. The calculated curves do fulfill the critical condition $(\partial P/\partial \rho)_c = 0$ not at the observed critical point but slightly above it. The deviation has also been noted in the case of Eq. (6.26), but the calculated densities of the coexisting phases meet at the

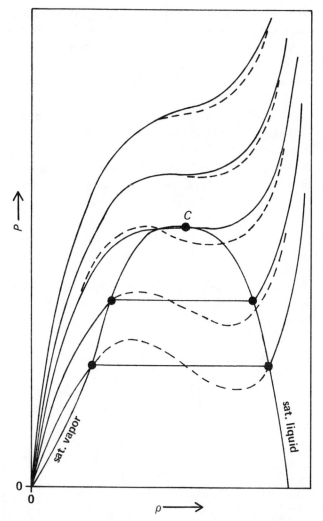

FIG. 7.11.   The pressure-density isotherms of a pure fluid. P and ρ are in arbitrary units, and the size of the critical region is exaggerated. The solid and dashed curves represent, respectively, the observed and the calculated data from an analytical equation of state.

observed critical point because it was enforced in the evaluation of the characteristic constants of pure fluids. The consequence of the fact that analytical equations yield $(\partial P/\partial \rho)_T < 0$ at $\rho = \rho^c$ and $T = T^c$ is that the pressures calculated at higher densities are too small over a range of temperatures below and above T^c. The errors in these derivatives cancel to some extent in mixtures when the critical locus is calculated relative to the pure components.

Systematic studies of the deviations from the analytical behavior began with the work of Guggenheim (1945), who showed empirically that the liquid-vapor coexistence curves $\rho(T)$ are cubic. Both the liquid branch $(\rho - \rho^c)$ and the vapor branch $(\rho' - \rho^c)$ are approximately proportional to $(T^c - T)^{1/3}$. Since then, it appears to be convenient to study the rates of change of various properties $Y(T, V)$ as functions of $\Delta\rho = (\rho - \rho^c)/\rho^c$ and $\Delta T = (T - T^c)/T^c$ on logarithmic plots. A *critical-point exponent* or *index* is defined as

$$\chi^{\pm} = \lim_{(T \to T^c)^{\pm}} [\ln Y(T, V)/\ln|\Delta T|] \tag{7.12}$$

where the plus sign indicates approaching from above T^c and the minus sign from below T^c. Four basic indices are distinguished, α, β, γ, and δ, for the limiting behavior of the following functions:

$$C_v \sim |\Delta T|^{-\alpha} \qquad (\rho = \rho^c); \tag{7.13}$$

$$\Delta\rho \sim |\Delta T|^{\beta} \qquad \text{(coexistence curve)}; \tag{7.14}$$

$$(\partial P/\partial \rho)_T \sim |\Delta T|^{\gamma} \qquad (\rho = \rho^c); \tag{7.15}$$

$$|(P - P^c)/P^c| = \Delta P \sim |\Delta\rho|^{\delta} \qquad \text{(critical isotherm)} \tag{7.16}$$

The values of the indices were established by taking into consideration all the known factors that might distort the equilibrium curves in the critical region, as well as the experimental errors of the measured variables. These efforts are outlined in the excellent reviews by Levelt-Sengers (1970, 1975), where the following values are given:

Index	Classical	Real fluid
α^+, α^-	0	0.07 ± 0.05
β	0.5	0.35 ± 0.01
γ^+, γ^-	1	1.22 ± 0.04
δ	3	4.4 ± 0.2

The "classical" values are those obtained from any equation of state by a Taylor expansion around the critical point and by applying the conditions of a critical point, (7.8) and (7.9). The real values are different; hence, the behavior of fluids in the critical region is called "nonanalytical." However, A, U, and P remain continuous functions of T and V. Rushbrooke (1963) and Griffiths (1965) derived from the stability conditions important inequalities between the critical indices:

$$\alpha^- + \beta(\delta + 1) \geq 2 \tag{7.17}$$

and

$$\alpha^- + 2\beta + \gamma^- \geq 2 \tag{7.18}$$

Widom (1965) introduced the so-called *scaling hypothesis*, according to which the difference between the chemical potential $\mu(T, \rho)$ and that on the critical isochore, $\mu(T, \rho^c)$, at the same T, is an antisymmetric function of a combination of $\Delta\rho$ and ΔT. Several important results are obtained from this hypothesis. One is that the inequalities (7.17) and (7.18) are equalities. Some empirical extrapolations up to the critical point are not confirmed; for example, the "law" of *rectilinear diameters*, discovered by Cailletet and Mathias in 1886:

$$(\rho + \rho')/2 = \rho^c + a_1|\Delta T| \tag{7.19}$$

where a_1 is a constant. As reviewed by Levelt-Sengers (1975), the result obtained from the scaling equations is rather

$$(\rho + \rho')/2 = \rho^c + a_2|\Delta T|^{1-\alpha} \tag{7.20}$$

The diameter, then, exhibits a slight curvature in the nearest vicinity of the critical point, and ρ^c obtained from (7.20) is smaller than that resulting from the equation of Cailletet and Mathias. Most of the known values of ρ^c were obtained by extrapolating the measured densities by means of (7.19) to T^c, but the resulting errors are small.

Hall and Eubank (1976) found that a relation analogous to (7.19) holds for the quantity $\Psi = (T^c/P^c)(\partial P/\partial T)_\rho$, which they called the *isochoric slope*:

$$(\Psi + \Psi')/2 = \Psi^c + a_3|\Delta T| \tag{7.21}$$

The value at the critical point, Ψ^c, is an important parameter in vapor pressure equations and certain equations of state.

There exist reliable vapor pressure equations that permit a long *interpolation.* If T^c and P^c are known and a section of the vapor pressure curve—say, up to the normal boiling point—is known, the values in the large gap up to P^c may be predicted with confidence. On the contrary, none of the equations for the densities of the phases $\rho_\sigma(T)$ is adequate for such an interpolation even when ρ^c and the low-temperature data are known. Recently, Thompson (1979) derived an accurate relation for $\rho_\sigma^*(P/P^c)$ that permits prediction of the densities of saturated liquid and vapor with, respectively, an accuracy of 0.4 percent and 0.9 percent. Small corrections must be introduced into the observed values of P^c and ρ^c, possibly in accordance with Eq. (7.20), to have a common relation for the liquid and vapor density curves.

Many papers have been published on the extensions and the applications of the scaling hypothesis. They are reviewed by Vincentini-Missoni (1972), Stanley (1971), Stephenson (1971), Chapela and Rowlinson (1974), and Levelt-Sengers (1975).

Errors in the thermodynamic functions calculated from an accurate equation of state, due to the nonanalytical behavior of fluids in the critical region, are generally small. The error in P along the critical isotherm, shown in Fig. 7.11, is usually less than 0.2 bar. Chapela and Rowlinson (1974) calculated the heat capacities C_v of carbon dioxide from an accurate multiparameter equation and from the scaling equations. The results are compared in Fig. 7.12. The value of C_v is finite at the critical point, whereas C_p, related to C_v by (1.32), tends to infinity. The residual heat capacity is related by thermodynamics to the pressure derivative

$$C_v^r = -T \int_0^\rho \left(\frac{\partial^2 P}{\partial T^2} \right)_\rho \frac{d\rho}{\rho} \tag{7.22}$$

Starting with (7.22), Goodwin (1975, 1978) developed an accurate multiparameter equation of state that incorporates the nonanalytic behavior in the critical region. Douslin and Harrison (1976) measured the vapor pressures and the *PVT* relations of ethylene and found that the second derivative along the saturation curve $(\partial^2 P/\partial T^2)_\sigma$—not at constant ρ—is finite at T^c, whereas according to the scaling equations, it should tend to infinity. Since the precision of the measurements in Douslin's laboratory is the highest possible with contemporary equipment, the nonanalytical behavior must be limited to an extremely nar-

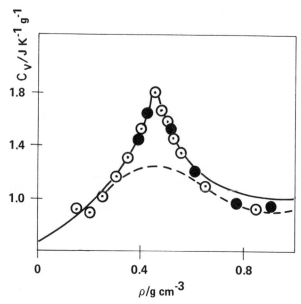

FIG. 7.12.   The heat capacity of carbon dioxide at constant volume and $T = 305.15$ K ($T^c = 304.21$ K). The dashed line was calculated from an equation of state, accurate outside the critical region; the solid lines were calculated from the limiting scaling equations. The experimental points were determined by Kruger (open circles) and Amirkhanov et al. (filled circles). The references are given by Chapela and Rowlinson (1974). The solid line has been corrected to eliminate an error, noted by Rowlinson, in the original paper.

row temperature range as far as the derivatives $(\partial P/\partial T)_\sigma$, $(\partial^2 P/\partial T^2)_\sigma$, and $(\partial H/\partial T)_{\rho^c}$ are concerned.

The scaling equations are valid only in the limits of small $\Delta\rho$ and ΔT. Doubtless, they would complicate greatly the relations for μ_i of the components. Fortunately, the results given above show that the deviations due to nonanalytical behavior of fluids are not serious. They will have even less influence in mixtures whose properties are usually calculated relative to those of the pure components.

The scaling equations for mixtures were derived by Griffiths and Wheeler (1970) and are reviewed together with the experimental evidence by R. L. Scott (1972, 1978). The difference in composition of the liquid and the vapor phase, as the equilibrium isotherm approaches the critical point, vanishes asymptotically as

$$|x - x'| \sim |\Delta P|^\beta \qquad \text{(const. } T) \qquad\qquad (7.23)$$

except, as shown by Andersen and Wheeler (1979), in the special case when the line of azeotropic points meets the critical curve. A relation analogous to (7.20) exists for the miscibility curve of two liquids with $(\rho + \rho')/2$ replaced by $(x_1 + x_1')/2$. Kreglewski (1963) found that the rectilinear diameter law holds only when the critical composition in a binary system x_i^c is close to 0.5. Otherwise, a slight curvature of the $(x_1 + x_1')/2$ line is noted just below T^c, the deviation going away from $x = 0.5$. These results are given in Table 7.1. Nakata et al. (1982) determined the indices near the LCSP of a binary system and obtained $(1 - \alpha) = 0.65$ and $\beta = 0.332 \pm 0.002$.

Among the most important thermodynamic functions for mixtures, only H^E, C_p^E, and the coefficient of thermal expansion are significantly affected in the vicinity of the critical solution points. These distortions are considered by Gaw and Scott (1971), Klein and Woermann (1975), Greer and Hocken (1975), Morrison and Knobler (1976), and Thoen and his coworkers (1978). As noted by Rowlinson (1969), the consequence of the fact that $\gamma^+ > 1$ is that the UCST calculated from an analytical equation of state with the parameters chosen to fit G^E outside the critical region (homogeneous liquid mixture) will always be too high, whereas the LCST will be too low. The calculated liquid-liquid critical locus is then expected to run in the (P, T, x_i) space slightly outside the observed curve.

The scaling hypothesis is an abstract thermodynamic concept, remote from a physical model of critical phenomena. The physical state is that of equilibrium between the clusters of molecules of various sizes. Two phases in the thermodynamic sense do not exist here. A very narrow range of cluster size distributions corresponds to a critical point. Small changes in $\Delta\rho$ or ΔT produce a large disturbance of the

TABLE 7.1.   Deviations from the rectilinear diameter of miscibility curves.

System		T^{ucs}	x_i^c	x_i^c
(1)	(2)	°C	observed	rectilinear
$n\text{-}C_8H_{18}$ +	$n\text{-}C_8F_{18}$	75.46	0.645	0.638
$n\text{-}C_8H_{18}$ +	CF_3COOH	57.79	0.343	0.355
$n\text{-}C_8H_{18}$ +	$(CF_3CO)_2O$	45.62	0.466	0.466

size distribution. The first statistical theory of clustering is due to Mayer (1938, 1940), who obtained the results that the coexistence curve has a flat top and the $P(\rho)$ isotherms contain a horizontal section at and just above the critical point (in fact, they are very nearly horizontal).

Stell (1969) and Mandel (1973) studied the radial distribution functions $g(r)$ in the critical region, obtained from the Percus-Yevick theory of fluids. Far away from the critical point the function varies with r^{-6}:

$$g(r) - 1 \approx \text{const. } r^{-6} \tag{7.24}$$

Near the critical point, it varies with r much more slowly:

$$g(r) - 1 \sim (4\pi\rho_n\lambda)^{-1} \exp[-(\lambda kT)^{-1}(\partial P/\partial\rho_n)_T^{1/2}] \tag{7.25}$$

where ρ_n is the number density, and

$$\lambda = (2\pi\rho_n/3) \int_0^\infty r^4 c(r) \, dr$$

where $c(r)$ is the *direct correlation function*, in the Percus-Yevick theory,

$$c(r) = g(r)\{1 - \exp[u(r)/kT]\} \tag{7.26}$$

The value of λ varies slowly with r, and at the critical point, $g(r) - 1 \sim (4\pi\rho_n\lambda)^{-1}$. While $g(r) - 1$ given by (7.24) can be neglected beyond the distances $r \approx 3\sigma$, $g(r) - 1$ given by (7.25) extends up to larger values of r and dominates at and near the critical point. Hence, *the Helmholtz energy A^r/RT should contain an exponential function of $\Delta\rho$ and $|\Delta T|^{-1}$*. By differentiation with respect to T, the function appears again in U^r/RT and C_v^r/R. The correction to C_v^r, obtained from scaling equations and shown in Fig. 7.12, may well be expressed by an exponential term.

A cluster in Mayer's theory was a "mathematical cluster." Several theories of *physical clusters*, or supermolecules with variable numbers of molecules, have been advanced, a rigorous one by Gillis, Marvin, and Reiss (1977), who also discuss earlier theories in detail. Here, the *translational* partition function of the vapor is replaced by that for a mixture of hard spheres of various sizes. The diameters of these spheres σ_s are those of the clusters of s molecules most probable under the given conditions. There may exist a number of "isomers" with the same s but different σ_s. For large s, a single isomer appears to dominate the distribution strongly. The problem was solved exactly in only one dimension (a row of hard rods with length l_s) with pair interactions given by the square-well potential, but the work may lead to an approximate theory in three dimensions.

Mruzik, Abraham, and Pound (1978) carried out molecular dynamics simulations of the nucleation process (formation of droplets) and made interesting comparisons with recent theories.

B. Problems in the Calculation of Phase Equilibria at High Pressures

As illustrated by Figs. 7.2 and 7.3, mixtures seem to differ qualitatively from pure fluids in their behavior in the critical region. The functions considered here are $P(T, x)$ and $\rho_m(T, x)$. Griffiths and Wheeler (1970) and Leung and Griffiths (1973) suggested using the chemical potentials as the *independent* variables—for example, for a binary system $P(\mu_1, \mu_2, T)$. This function contains intensive thermodynamic quantities only (called *fields* by Griffiths and Wheeler), which are always the same in two coexisting phases: μ_1 tends to $-\infty$ when $x_1 \to 0$, and μ_2 tends to $-\infty$ when $x_1 \to 1$. In order to obtain variables which are finite everywhere along the critical locus curve, Leung and Griffiths introduced the variables

$$\zeta_1 = \frac{C_1 \exp(\mu_1/RT)}{C_1 \exp(\mu_1/RT) + C_2 \exp(\mu_2/RT)} \tag{7.27}$$

and $\zeta_2 = 1 - \zeta_1$. Here C_1 and C_2 are positive constants; ζ_1 is zero at $x_1 = 0$ and equals unity at $x_1 = 1$. The quantities ζ_i are not known, nor can they be directly measured. However, by using proper transformations, Leung and Griffiths expressed the molar densities ρ_m and the mole fractions as functions of ζ_i. The constants C_1 and C_2 can then be found by an iteration from the $P(\rho_m, T)$ data for the pure components. The advantages of the concept are clearly shown on models made by Moldover and Gallagher (1977), one of which is schematically shown in Fig. 7.13. It is seen that the dashed curve for a mixture at constant ζ_i has, qualitatively, the same shape as the curves for the pure compo-

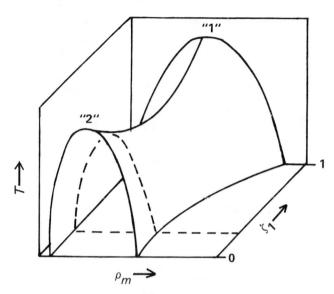

FIG. 7.13.   Liquid-vapor equilibrium curves near the critical locus curve. The critical points of the components are marked by "1" and "2." The dashed curve is for mixtures at constant ζ_1. The critical locus curve along the top of the saddle is that of a system exhibiting a minimum on the $T^c(x)$ curve (a special case, of course).

nents. *The humps*, corresponding to the retrograde condensation and *appearing at constant x_i* in Fig. 7.3, *disappeared*. Therefore, Moldover and Gallagher operated with the functions $T^c(\zeta_i)$ and $\rho^c_m(\zeta_i)$ and the same functions for the pure components and the mixtures: say, the rectilinear diameter law expressed in terms of ζ_i. If the critical locus was determined experimentally—that is, $T^c(x)$, $P^c(x)$, and $V^c_m(x)$ are known—$T^c(\zeta)$ and $\rho^c_m(\zeta)$ were found by an iteration. Next, the densities and the pressures of the coexisting phases were obtained at constant T. In this way, accurate liquid-vapor equilibrium isotherms were obtained down to temperatures well below the critical state for the CO_2 + C_2H_6 and SF_6 + C_3H_8 systems. Leung and Griffiths made earlier similar calculations for the ^3He + ^4He system. We note that the critical locus curves must be known in advance. It is, then, not a method for predicting phase equilibria but an interpolation procedure.

The curves for the pure components and the mixture shown in Fig. 7.13 have the same shape and differ only by scale factors. These curves, if plotted by using the reduced quantities T/T^c and ρ_m/ρ^c_m, would be identical at least at and in the vicinity of the critical points. Hence, the variable ζ_i puts different systems into corresponding states.

It is clear that the critical conditions, expressed by any of the accurate equations of state, are complicated equations. For this reason a simple two-parameter equation is usually chosen for this purpose. The most popular is the Redlich-Kwong (1949) equation

$$P = \frac{RT}{V_m - b} - \frac{a}{T^{1/2}V_m(V_m + b)} \tag{7.28}$$

where a and b are the characteristic constants. Their values, calculated from the critical point conditions, are

$$a = Z^c(1 + b_o)\left[1/(1 - b_o) - Z^c\right] R^2(T^c)^{2.5}/P^c$$

and

$$b = b_o V^c = b_o Z^c RT^c/P^c$$

where $b_o = 0.260$ and Z^c is given by Eq. (6.54).

Gas-liquid critical locus curves were calculated, using this equation, by Joffe and Zudkevitch (1967), Spear, Robinson, and Chao (1969), and Hissong and Kay (1970). Since the first term in (7.28) is valid only at $\rho \to 0$, the parameter b obtained from the data at $\rho = \rho^c$ is not a pure function of σ^3 (as was supposed), and parameter a is not necessarily proportional to $\bar{u}\sigma^3$. Hence, the combining rules for a_{12} and b_{12} cannot be applied with any confidence. In fact, Huron, Dufour, and Vidal (1977–78) calculated the values of a_{12} and b_{12} fitting the critical locus curves of CO_2 + alkane and H_2S + alkane systems, and found that they are different from those required to fit liquid-vapor equilibria. This example shows that the less convenient but accurate enough equations of state with a good theoretical foundation should be used in any case.

There are only a few examples of such calculations. Teja and Rowlinson (1973) applied the concept of shape factors, the van der Waals approximation for $\bar{u}_m$ (Chap. 4), and the combining rule (5.6) to the evaluation of the gas-liquid $P^c(x)$ and $T^c(x)$ curves with good results. The curve calculated for the CO_2 + $n\text{-}C_8H_{18}$ system, with $k^u_{12} = 0.7$, exhibits the gas-gas critical locus of the second type.

Neff and McQuarrie (1975) compared various theories with respect to their capability of correctly predicting gas-gas equilibria and found that the van der Waals approximation yields the best results. Two other approximations, known as the two-fluid and three-fluid concepts (and considered also by Breedveld and Prausnitz, 1973), proved to be unsatisfactory. For this and other reasons, the two approximations are not considered in this book.

Twu, Gubbins, and Gray (1976) and Gubbins and Twu (1978) calculated the liquid-liquid, gas-liquid, and gas-gas equilibria for the model systems of argon + krypton mixtures, adding to either argon or to krypton a dipole or a quadrupole moment of varying strength. They were able to obtain all the major types of phase diagrams. An example is shown in Fig. 7.14, where the quadrupolar contributions are added to the Helmholtz energy of krypton, leaving argon as it is. The curves exhibit the transition from a continuous critical curve for $Q_{Kr}^* = 0$ to

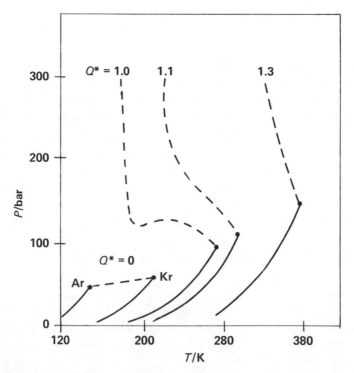

Fig. 7.14.   The variation of the critical locus curves of the argon + krypton system (dashed) with the reduced moment $Q^* = Q/(\bar{u}\sigma^5)^{1/2}$ added to krypton. The solid lines are the vapor pressure curves of the pure components.

that of the second type, and finally to the gas-gas curve of the first type. There appears also a limited miscibility of the liquid phases at lower temperatures when $Q_{kr}^* \geq 1$, not shown in the figure. The contribution of Q to $\bar{u}_{12}$ is zero in all the cases because it is proportional to $Q_1 Q_2$, while Q_1 (argon) is zero. Thus, the weakness of $\bar{u}_{12}$ compared to $\bar{u}_{22}$ is enhanced by Q_2.

The effect of size difference among the components has not yet been so accurately studied. Neff and McQuarrie found that with increasing σ_{22}/σ_{11} at constant $\bar{u}_{22}/\bar{u}_{11}$, the system tends from the second type to the first type of gas-gas equilibria; that is, the region of immiscibility broadens.

Both Teja and Rowlinson, and Gubbins and Twu used accurate multiparameter equations of state for the reference substances. These equations do not contain a separate repulsion term, although they take proper care of the intermolecular repulsion. It is then interesting to see the results obtained with an equation of state containing such a term. McGlashan, Stead, and Warr (1977) applied Eq. (6.18) to the problem of gas-gas immiscibility. The results obtained for the He + CH$_4$ system, with the value $k_{ij}^u = 0.8$ in rule (5.6), fitting well the gas-gas critical locus curve, are qualitatively good, but the miscibility gaps are narrower than the observed wide gaps. This indicates that the attraction term is slightly too large compared with the (accurate) repulsion term. The value of k_{ij}^u in Eq. (5.6) is close to that obtained for an accurate pair potential for the He + Ar system (Table 5.1).

For the purposes of the discussions in this and the next chapters, we summarize the basic relations for $\bar{u}_m$:

$$\bar{u}_m = (V_m^o)^{-\delta} \left(\sum_{h=1}^{m} x_h g_h^2 \right) \sum_i \sum_j \varphi_i \varphi_j \, \varepsilon_{ij} (V_{ij}^o)^\delta \qquad (7.29)$$

The power δ, introduced by Teja, allows comparison of different approximations for $\bar{u}_m$. In the van der Waals approximation, $\delta = 1$ and all $g_i = 1$. For $\delta = 0$, Eq. (7.29) simplifies to the square-well approximation where either all $g_i = 1$ (mole fractions) or $g_i = (V_i^{oo}/V_1^{oo})^{1/3}$, $(i = 1, 2, \ldots)$; that is, surface fractions, and $\varepsilon_{ii} = \bar{u}_{ii}/g_i^2$ will be used:

$$V_m^o = \sum_i \sum_j x_i x_j V_{ij}^o; \qquad (7.30)$$

$$V_{ij}^o = k_{ij}(V_i^o + V_j^o)/2; \tag{7.31}$$

$$\varepsilon_{ij} = \varepsilon_{ij}^o \left[1 + \frac{\eta_{ij}}{kT} + \left(\frac{\delta_{ij}}{kT} \right)^2 \right]; \tag{7.32}$$

$$\varepsilon_{ij}^o = 2\gamma_{ij} \left(\frac{1}{\varepsilon_{ii}^o} + \frac{1}{\varepsilon_{jj}^o} \right)^{-1} \tag{7.33}$$

where k_{ij}, γ_{ij}, η_{ij}/k, and δ_{ij}/k are constants for a given system. The term η_{ij}/k will be assumed equal to $(\eta_{ii} + \eta_{jj})/k$. Large deviations from the random mixing approximation may to some extent be corrected by making k_{ij} a function of x_i (Kay and Kreglewski, 1983). For pure fluids, Eq. (7.29) becomes

$$\bar{u}/k = (\bar{u}^o/k)(1 + \eta/kT) \tag{7.34}$$

because no attempts have been made to determine δ_{ii}/k for pure fluids.

It is known that as the ratio $\bar{u}_{22}/\bar{u}_{11}$ increases, the calculations of the phase diagrams of mixtures become more difficult, and the extrapolations to higher pressures become less reliable. Particularly, the calculated gas-liquid critical pressures P^c are always too high, whereas the gas-gas P^c values (if they appear) are always too low, as shown schematically by the dashed curves A in Fig. 7.15. The full curves and the critical points C_1 (gas-liquid) and C_2 (gas-gas) are the observed values for a binary system. If the interaction constants for $\bar{u}_{ij}$ and V_{ij}^o are fitted to the observed critical point C_2, the calculated gas-gas equilibrium curves B run inside the observed curves. These deviations occur even if the most accurate of known equations of state are used. The deviations may be partly eliminated by taking $k_{ij} > 1$ in Eq. (7.31). The remaining deviations are much too large to be corrected by means of the scaling equations, which are valid only for small $\Delta\rho$. At the present time, we know only that μ_i should contain a negative correction term. Eq. (1.39) for μ_i is exact but the terms A_m^r and $(\partial A_m^r/\partial x_j)_{x_k}$ calculated for a random mixture may be inaccurate. The correction term may be related to μ_i^{or}, derived from Eq. (5.27),

$$\frac{\mu_i}{RT} = \ln x_i + \frac{A_m^r}{RT} + Z_m - 1 + \ln \rho_m$$

$$- \sum_{j \neq i} x_j \left[\frac{\partial(A_m^r/RT)}{\partial x_j} \right]_{T, \rho, x_k} + \frac{\Delta\mu_i^{or}}{RT}; \quad (k \neq i, j) \tag{7.35}$$

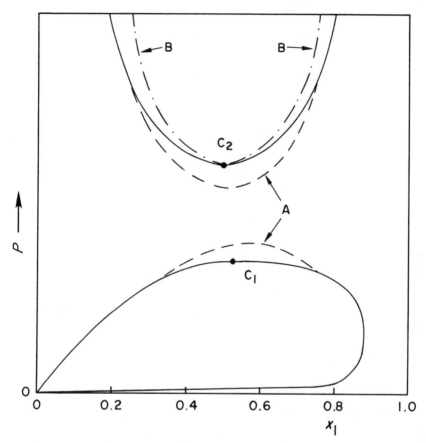

FIG. 7.15.   Gas-liquid and gas-gas $P(x)$ isotherms of binary systems with $\bar{u}_{22}$ $\gg \bar{u}_{11}$. The dashed curves A show typical deviations of calculated values from the observed values (full curves) near the critical points C_1 or C_2. Curves B are discussed in the text.

where A_m^r and Z_m are those for a random mixture (Eq. 7.29) calculated by using an equation of state that yields reasonably accurate saturation densities up to the critical points of the components. A^{or} in Eq. (5.27) for a liquid mixture at low temperatures is not known, and it is usually neglected when the interaction parameters are fitted to the experimental data. As the pressure increases at a constant T, the densities of the coexisting phases, ξ_m and ξ'_m, approach each other, and the ordering effects in the liquid phase decrease; however, the calculated values of ξ_m and ξ_m' approach each other less rapidly and meet at a higher pressure than the observed P^c because the "decrease of order" in the liquid (and a possible increase in the vapor) has not been taken into account. Kreglewski and Hall (1983) estimated the effect of the "change of order" in the liquid relative to that in the vapor by means of the relations

$$\Delta A^{or}/RT = -(B_1/T)x_i x_j \exp[-B_2(\xi_m - \xi_m')_\sigma^2] \tag{7.36}$$

and

$$\Delta \mu_i^{or}/RT = -(B_1/T)\, x_j^2 \exp[-B_2(\xi_m - \xi_m')_\sigma^2] \tag{7.37}$$

They have found empirically that B_1 varies with $\Delta T^c = (T_2^c - T_1^c)/(T_2^c + T_1^c)$ rather than with $(\Delta T^c)^2$, as Eq. (5.27) suggests:

$$B_1 = B_o|\Delta T^c| \tag{7.38}$$

where the values of the constants obtained for several systems are $B_o \approx 110 \pm 40$ and $B_2 \approx 20 \pm 5$. Although the exponential function appears in the theory of fluctuations, the foregoing relations are nothing more than empirical corrections. If we neglect $(\partial \xi/\partial T)_\sigma$ and $(\partial \xi'/\partial T)_\sigma$, the corrections for the residual internal energy, ΔU^{or}, and the entropy, ΔS^{or}, are

$$\Delta U^{or} \approx \Delta A^{or} \quad \text{and} \quad \Delta S^{or} \approx 0.$$

Phase equilibria in a binary system are calculated as follows. The mole fractions x_1, x_1', and x_1'' and the molar densities ρ_m, ρ_m', and ρ_m'' of the coexisting phases are obtained by iteration at the given T and P, using the Gibbs equilibrium conditions and Eq. (7.35) for the chemical potential. We choose an isotherm T_1 for which reliable bubble- and dew-point pressures and densities up to the critical point are available. The densities are most sensitive to the value of k_{12}, whereas the saturation pressures depend mostly on γ_{12} and k_{12}. The values of these parameters are fitted to the data far from the gas-liquid critical point, initially keeping $\delta_{12}/k = 0$ (Eq. 7.32), $B_o = 100$, and $B_2 = 20$ (Eq. 7.37). Next, the values of P^c and x_1^c are calculated as the maximum point on the $P(x)$ isotherm at the temperature T_1—as explained in the discussion following the condition stated as Eq. (7.10)—varying the value of B_o to fit as well as possible the observed gas-liquid critical point. The procedure is repeated for another isotherm T_2 to find whether γ_{12} is the same as for T_1. If it is the same, $\delta_{12}/k = 0$. If the values of γ_{12} obtained

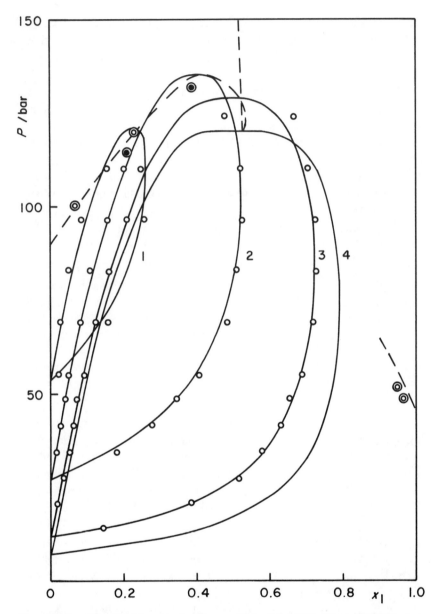

FIG. 7.16.   The liquid-vapor isotherms $P(x)$ of the CH_4 (1) + H_2S (2) system. Experimental data of Kohn and Kurata (open circles) and Reamer et al. (filled circles). The double circles are P^c. The curves in Figs. 7.16 to 7.20 were calculated at T/K: 344.26 (1); 310.93 (2); 277.59 (3); 263.15 (4); 243.15 (5); 233.15 (6); 223.15 (7); 213.15 (8); 198.70 (9); 193.15 (10). The dashed curves are $P^c(x)$.

for two or three isotherms differ, they are plotted against $1/T^2$. Extrapolation to $1/T^2 = 0$ yields the proper value of γ_{12}, and δ_{12}/k is then calculated from (7.32). The liquid-liquid coexistence curve is very sensitive to the values of γ_{12} and δ_{12}/k, and they should be accordingly checked if phase separation occurs in the system.

The results obtained by Kreglewski and Hall (1983) for the methane + hydrogen sulfide system are shown in Figs. 7.16 to 7.20.[†] Liquid-vapor equilibrium isotherms $P(x)$, the critical locus curves, and the densities of the saturated phases $\rho(x)$ were measured by Reamer et al. (1951) and by Kohn and Kurata (1958) at 182 to 364 K. Few data are reported by Robinson et al. (1959), who studied the CH_4 + CO_2 + H_2S system. Kohn and Kurata discovered the existence of two liquid phases and determined the quadruple point (182.2 K, 33.8 bar). They determined the $P(T)$ curves at constant $x_1 = 0.067, 0.229,$ $0.542, 0.752, 0.890,$ and 0.9475 and found that there are no gas-liquid critical points in the range $x_1 = 0.542$ through 0.890 (at pressures up to 131 bar). The data of Kohn and Kurata, shown in Fig. 7.16, agree very well with those of Reamer et al. except in the critical region. The curves shown in the figures were calculated by means of the equation of state (6.26), using the square-well approximation (Eq. 7.29; $\delta = 0$, surface fractions), the characteristic constants of the pure components from Table 6.4, and the following mixed interaction parameters:

$$\gamma_{12} = 0.94; \quad \eta_{12}/k = 8.0 \text{ K}; \quad \delta_{12}/k = 54.4 \text{ K}; \quad k_{12} = 1.10$$

and the constants of Eq. (7.37), $B_o = 73 \, (\pm 10)$ and $B_2 = 20 \, (\Delta T^c$ $= 0.3239)$. The calculated $P^c(x)$ curve agrees with the data of Kohn and Kurata and points to the absence of critical points for $x_1 > 0.53$ up to about 0.90. The critical curve appears again from the UCEP $(x_1 = 0.90, T \approx 213 \text{ K})$ to the critical point of pure CH_4. The critical point D at 277.6 K (Figs. 7.18 and 7.19) was probably not well located by the graphical procedure employed by Reamer et al. (because of the flat top of this isotherm). Kohn and Kurata present the liquid-liquid equilibrium data on graphs only. The calculated results, shown for isotherms 8 and 9 in Fig. 7.17, qualitatively agree with these data.

After passing through a shallow minimum at about 243 K, the $P^c(x)$ curve turns up to meet the gas-gas or rather the fluid-fluid critical locus. Let us note that isotherms 8 and 10 in Figs. 7.20a and 7.20b

[†]Reproduced with permission from *Fluid Phase Equil.* (1983): 11.

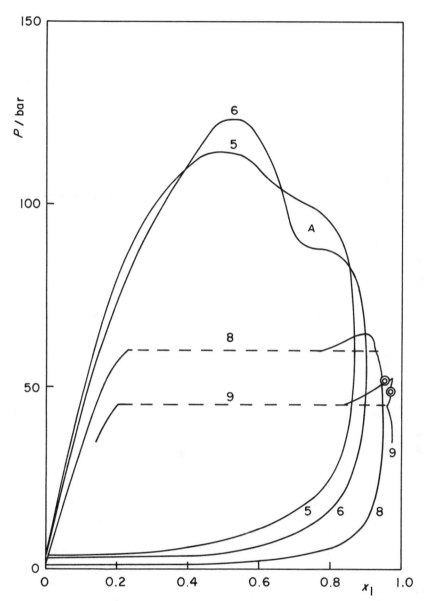

FIG. 7.17.   Liquid-liquid and liquid-fluid isotherms $P(x)$ of the CH_4 (1) + H_2S (2) system. The temperatures are given in Fig. 7.16.

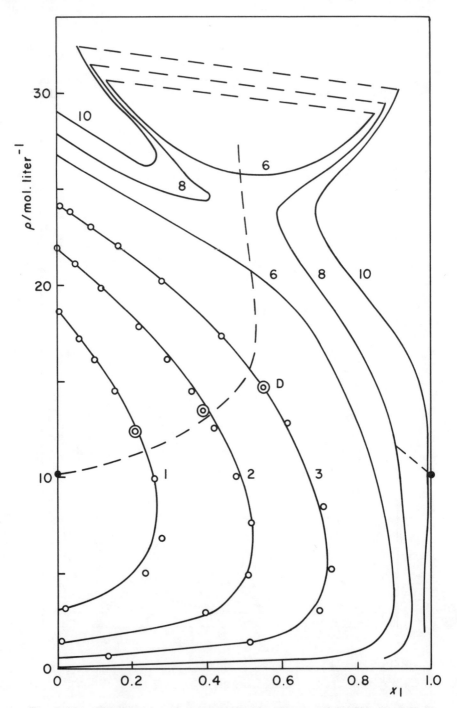

FIG. 7.18.   The densities of saturated liquid and vapor of the CH_4 (1) + H_2S
(2) system, from the experimental data of Reamer et al. The curves were calculated at T/K given in Fig. 7.16. The dashed curve is $\rho^c(x)$. The double
circles are ρ^c. (Point D is discussed in the text.) The dashed straight lines
connect the fluid phases at 3,000 bar.

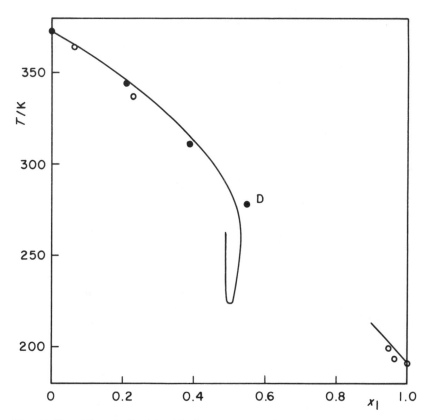

FIG. 7.19.   The gas-liquid critical temperature curve of the CH$_4$ (1) + H$_2$S (2) system: the calculated curves and the experimental points. (For references, see Fig. 7.16. Point D is discussed in the text.)

represent the effect of pressure on the liquid-liquid miscibility gap, whereas isotherms 4, 5, and 6 are what is usually called gas-gas equilibrium curves. This system appears to be an example of a continuous passage from the liquid-liquid to the gas-gas equilibrium. Isotherm 7 (223.15 K) appears to have no critical point and must be close to the double critical point temperature. The iterations seldom converge to the equilibrium values in the vicinity of this point, and the pressure at this point may be only roughly estimated.

As the ratio $\bar{u}_{22}/\bar{u}_{11}$ of the components increases, so does the value of k_{12} in Eq. (7.31) that is required to fit the liquid-vapor equilibrium pressures. Analogous results were obtained by Simnick et al.

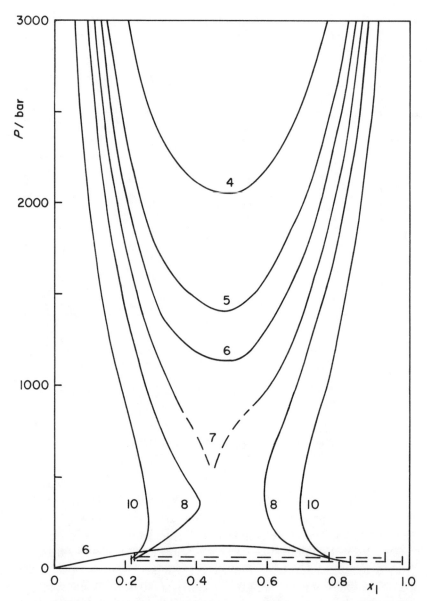

FIG. 7.20a.   The fluid-fluid equilibrium isotherms $P(x)$ predicted for the CH_4 (1) + H_2S (2) system at temperatures given under Fig. 7.16. The liquid-liquid-vapor equilibria are shown by the dashed lines.

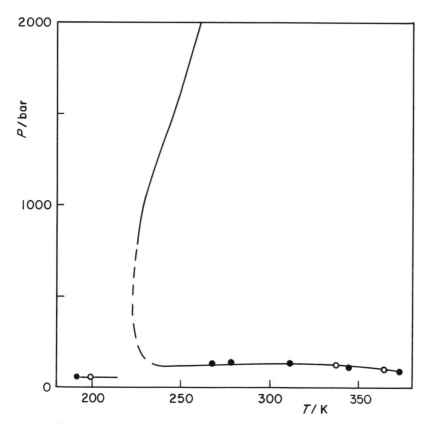

FIG. 7.20b.   The gas-liquid and fluid-fluid critical curve P^c vs. T^c of the CH_4 (1) + H_2S (2) system (for references, see Fig. 7.16). The double critical point appears at about 223 K.

(1979), using the same equation of state: the values of k_{ij} of the H_2 + n-C_4H_{10} and H_2 + C_6H_6 systems are abnormally large. One of the consequences is that the liquid densities, calculated for mixtures rich in the less volatile component, are too low. Good values obtained for the CH_4 + H_2S system are rather an exception to this rule.

The systems N_2 + CO_2 and CH_4 + CO_2 have been thoroughly investigated. By employing the techniques outlined above, Kreglewski and Hall (1983) obtained $\delta_{12}/k = 44.9$ K for the first system and $\delta_{12}/k = 0$ for the second system. In both cases the corrections ΔA^{or} and ΔU^{or} were introduced. These corrections are limited to the saturation curves and are zero for a compressed homogeneous fluid. Since ΔU^{or}

affects U_m^r and H_m^r, it is proper to check whether the absence of ΔU^{or} in the relations for a compressed gas mixture *just above the critical point* does not adversely affect the values of H^E, which is also sensitive to the value of δ_{ij}/k. The results obtained are compared in Fig. 7.21 with the data of Lee and Mather (1972) and Hejmadi et al. (1971). The value of H^E was calculated from H_m^r/RT of the mixture (Eq. 6.33) by subtracting $\Sigma x_i H_i^r/RT$ at the same T and P. At 313 K, the systems are only 9°C above T^c of CO_2, and at $P = 101$ bar, the densities are close to the critical densities of the systems. The agreement with the data is satisfactory. Nevertheless, Eq. (7.36) introduces a discontinuity of the thermodynamic properties near the critical points of mixtures, which—however small—causes a problem.

The passage from a saturated to an unsaturated state can be made continuous as follows. Since we may approach the critical state of a system at a constant temperature by varying x_i or ξ_m or both, the correction $\Delta\mu_i^{vc}$ (superscript *vc* for vicinity of critical point) along the saturation curve may be expressed by the empirical relation

$$\Delta\mu_i^{vc}/RT = - (b_1/T)\ x_j^2\ \exp[-b_2(\xi_m - \xi_m')^2 - b_3(x_j - x_j')^2] \tag{7.39}$$

where $b_2 \approx b_3$ and b_1 are (positive) constants. Once the critical locus, ξ_m^c and x_j^c, has been calculated for the given isotherm, the terms $(\xi_m - \xi_m')_\sigma^2$ and $(x_j - x_j')_\sigma^2$ in (7.39) may be replaced by $4(\xi_m - \xi_m^c)^2$ and $4(x_j - x_j^c)^2$, respectively, where ξ_m and x_j are the densities and compositions either along the saturation curve or beyond it—that is, for the homogeneous fluid.

When the critical temperatures of the components are similar or equal, $\Delta\mu_i^{or} \approx 0$. The gas-liquid critical locus of the $CO_2 + C_2H_6$ system calculated with $\Delta\mu_i^{or} = 0$ agrees very well with the observed values. The agreement is due to the accuracy of the equation of state: for most of the substances in Table 6.4 the densities of the saturated phases meet nearly exactly at the observed values of T^c and P^c (Fig. 6.5). This agreement extends to mixtures, and the corrections due to nonanalytical behavior of fluids can be ignored. The system $CO_2 + C_2H_6$ exhibits positive azeotropes caused by the inequality of the quadrupole moments of the components; however, the liquid phase remains homogeneous at all temperatures at which the system was examined. With increasing inequality of the multipolar forces, the sys-

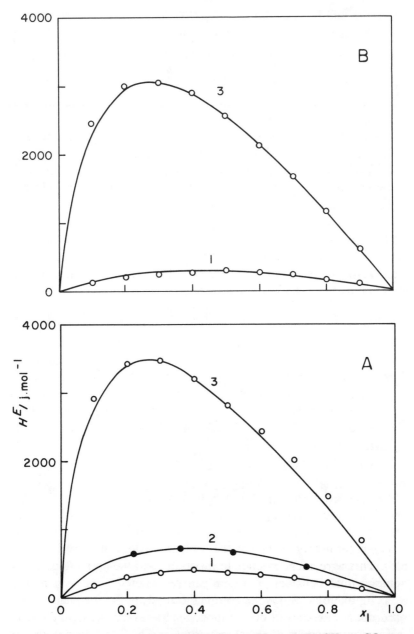

FIG. 7.21.   The excess enthalpy of (A) N_2 + CO_2 and (B) CH_4 + CO_2 gaseous mixtures at 313.15 K and P/bar: 50.66 (1); 65.50 (2); 101.33 (3). The curves were predicted. The experimental data are those of Lee and Mather (open circles) and Hejmadi et al. (filled circles). Reproduced with permission from *Fluid Phase Equil.* 1983, 15:11.

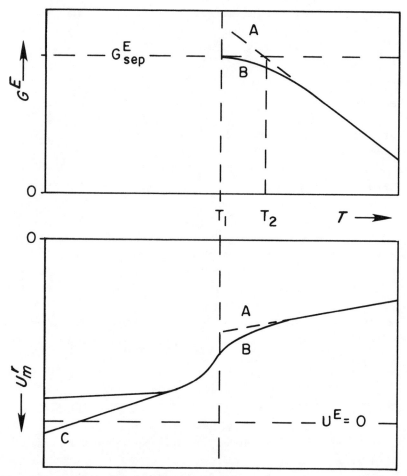

FIG. 7.22.   Temperature dependence of U_m^r and G^E at constant composition x_1 in the vicinity of T^{ucs} and x_1^{ucs}. G_{sep}^E is the value of G^E at which liquid-liquid phase separation may occur.

tem may tend to form two liquid phases at low temperatures—as, for example, mixtures of alkanes with perfluoro-alkanes. We may choose two components with similar gas-liquid T^c, so that $\Delta\mu_i^{gr} = 0$. As we approach T^{ucs} from above, U_m^r of the mixture is less negative than $\sum_i x_i U_i^r$, so that $U^E > 0$ (Eq. 2.26). G^E increases according to (2.19) until it reaches G_{sep}^E, at which liquid-liquid separation may occur, as shown schematically in Fig. 7.22. The values of G^E and U_m^r calculated for a random mixture at constant composition $x_1 = x_1^{ucs}$ will vary as shown by curves A, and phase separation will occur at T_2. However, as T^{ucs} is approached, molecular clusters of two kinds begin to form: those rich in component 1 and those rich in component 2. This causes a decrease of U_m^r along curve B by a nonrandom contribution U^{NR}, which varies most rapidly at T^{ucs}, where $C_v^r = (\partial U^r / \partial T)_v$ is the largest (Fig. 7.12). The real $T^{ucs} = T_1$ is a few degrees lower than T_2. This conclusion concurs with that deduced from the scaling hypothesis (see the discussion following Eq. 7.23). The excess enthalpies in this temperature range are always lower than the values calculated for a random mixture. At temperatures below T^{ucs}, U^E may remain positive or it may follow curve C and change sign. In the latter case, the system will tend to form a closed miscibility loop with an LCSP at which thermodynamics requires that $H^E < 0$. The calculations of a miscibility gap can be simplified by keeping a constant P just above the vapor pressure at T^{ucs}, thus eliminating the vapor phase. The gradual transition from a random to a nonrandom mixture near the critical locus has not yet been studied on the basis of an accurate equation of state. Eq. (7.39) for $\Delta\mu_i^{yc}$ may be helpful in these calculations.

The deviations of the properties of a liquid mixture at high pressures from those of an ideal system can be estimated by means of Eq. (2.27)—that is, $\mu_i^E = \mu_i^r - G_i^r$, at the same T and P. This definition was derived from (2.16), which is not restricted to $P = 0$. We may choose for P the saturation pressure of the more volatile component (or of the positive azeotrope, if it appears). The results obtained for the $CH_4 + C_2H_6$ system, calculated using the same relations as for the $CH_4 + H_2S$ system, are compared in Table 7.2 with the data of Miller et al. (1977) at 180 K. The calculated $P^c(x)$ and $T^c(x)$ curves also agree well with the observed critical locus curves. Two sets of μ_i^r values for the liquid phase are given in Table 7.2: the first along the saturation pressure $P(x)$, the second at constant $P = 32.57$ bar (calculated saturation pressure of CH_4). From the second set we obtain $\mu_1^E/RT = 0.178$ and $\mu_2^E/RT = 0.100$ at $x = 0.5$ for the compressed liquid mixture.

C. The Pseudocritical States of Mixtures

The concept of pseudocritical constants was first introduced by Kay (1936) as the hypothetical critical constants of hydrocarbon mixtures that are linear functions of x_i. The idea was to estimate empirically the deviations from an ideal mixture.

TABLE 7.2. Residual chemical potentials of the components of the system CH_4 (1) + C_2H_6 (2) at 180 K.

P_σ/bar	Observed		Calculated		Saturation P_σ		$P = 32.57$ bar	
	x_1	x_1'	x_1	x_1'	μ_1'/RT	μ_2'/RT	μ_1^r/RT	μ_2^r/RT
0.75			0	0	—	-0.0266[a]	—	-3.667[a]
1.84	0.0327	0.5734	0.0325	0.588	2.880	-0.914	0.115	-3.676
3.13	0.0745	0.7493	0.072	0.758	2.330	-1.439	0.092	-3.675
4.71	0.1241	0.8335	0.121	0.839	1.898	-1.839	0.063	-3.672
5.67	0.1567	0.8633	0.151	0.867	1.698	-2.018	0.045	-3.669
9.53	0.2825	0.9213	0.278	0.9221	1.116	-2.504	-0.030	-3.648
15.67	0.4928	0.9551	0.494	0.9556	0.517	-2.903	-0.153	-3.570
21.28	0.6898	0.9712	0.697	0.9717	0.131	-3.050	-0.252	-3.421
24.82	0.8025	0.9792	0.814	0.9800	-0.054	-3.066	-0.296	-3.283
28.28	0.9002	0.9874	0.911	0.9882	-0.198	-3.042	-0.322	-3.116
32.57			1	1	-0.333[b]	—	-0.333[b]	—

(a) G_2^r/RT.
(b) G_1^r/RT. All values of μ_i^r are for the liquid at the *calculated* x_1.
NOTE: For the compressed liquid state, $B_1 = B_2 = 0$.

The name *pseudocritical constants* or *locus curves* of mixtures, $T_x^c(x)$, $P_x^c(x)$, and $V_x^c(x)$, is now used to describe states that are more accurately related to the parameters of the intermolecular energy, $\bar{u}_m$ and σ_m, or in the notation used in the principle of corresponding states, f_x and h_x.

If f_{ii} and h_{ii} or f_{jj} and h_{jj} locate the critical points of the pure components, then f_x and h_x determine the pseudocritical point of a mixture treated as a *pure conformal fluid*. It fulfills at this point the thermodynamic conditions for a pure fluid, (7.8) and (7.9). However, real mixtures with the *same* f_x and h_x have an *observed critical locus* $T^c(x)$, $P^c(x)$, and $V_m^c(x)$, which conforms to the conditions (7.1) and (7.2) and usually differs from the pseudocritical curve. The differences are schematically shown in Fig. 7.23 at a temperature between the critical temperatures of the pure components. Curve A represents the variation of the volumes of coexisting liquid V_m^L and vapor V_m^G with the composition x_i. Inside it is the boundary curve B of material instability, $(\partial^2 G/\partial x^2)_{T,P} = 0$, which is tangential to curve A at the critical point C. It was called by van der Waals the *spinodal curve*. The innermost is the hypothetical curve D along which the condition $(\partial P/\partial V)_{T,x} = 0$ is ful-

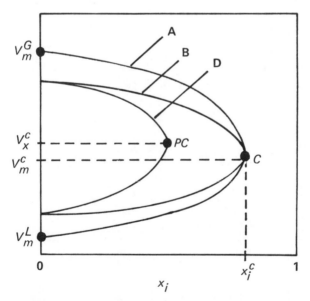

FIG. 7.23.   Schematic representation of the coexistence curve and the stability conditions on the $V_m(x_i)$ projection of the P-V_m-x_i surface.

filled, and PC is the pseudocritical point. It is also marked on the $P(T, x)$ projection in Fig. 7.1(C). It has been shown by Prigogine and Defay (1954) that for mixtures, curve D always lies inside the spinodal curve. The two become identical for the pure components.

If proper relations for f_x and h_x and the curves $T_x^c(x)$, $P_x^c(x)$, and $V_x^c(x)$ are known, the critical locus curves will be determined by the differences $(T^c - T_x^c)$, $(P^c - P_x^c)$, and $(V_m^c - V_x^c)$ at the given x_i. The basic relations for these differences were derived by Bellemans, Mathot, and Zuckerbrodt (1956) and by Byers Brown (1957) for the case of all the parameters $f_{ii}, \ldots, f_x$ and $h_{ii}, \ldots, h_x$ close to unity: that is, for molecules with similar values of $\bar{u}$ and σ. In this case the differences $(T^c - T_x^c)$, etc., are small. The treatment is best summarized by Rowlinson (1969), and the results are as follows. For a binary system,

$$T^c - T_x^c = - x_i x_j \left[\frac{(\partial P_x/\partial x_i)_{T, V}^2}{RT(\partial^2 P_x/\partial V\partial T)_x} \right]_c \tag{7.40}$$

where subscript c indicates the critical locus curve. The values of these derivatives can best be estimated from the simple equation of state (6.18). Kreglewski et al. (1973) found from the compressibility data for several gases that $\xi \approx 0.16\, V_m^c/V_m$, and the van der Waals "constant" $a^c \approx 1.7\, RT^c V_m^c$ (if treated as independent of T). The first of these coefficients is very close to the value 0.1655 obtained by Bienkowski and Chao (see Eq. 6.46). Hence, for mixtures, $\xi_m \approx 0.16\, V_x^c/V_m$, $a_m^c \approx 1.7\, RT_x^c/V_x^c$, and from (6.18),

$$\left(\frac{\partial P_x}{\partial x_i} \right)_{T, V} = \frac{0.16\, RT}{V_m^2} \left[\frac{4 + 4\xi_m - 2\xi_m^2}{(1 - \xi_m)^4} \right] \left(\frac{\partial V_x^c}{\partial x_i} \right)_c$$

$$- \frac{1.7\, R}{V_m^2} \left[V_x^c \left(\frac{\partial T_x^c}{\partial x_i} \right)_c + T_x^c \left(\frac{\partial V_x^c}{\partial x_i} \right)_c \right]; \tag{7.41a}$$

$$\left(\frac{\partial^2 P_x}{\partial V\partial T} \right)_x = - \frac{RF_x}{(V_x^c)^2} \tag{7.41b}$$

where F_x is a function of η/k. For $T = T_x^c$ and $V_m = V_x^c$, Eq. (7.41a) becomes

$$\left(\frac{\partial P_x}{\partial x_i} \right)_c = - \left[\frac{1.7}{T_x^c} \left(\frac{\partial T_x^c}{\partial x_i} \right)_c + \frac{0.22}{V_x^c} \left(\frac{\partial V_x^c}{\partial x_i} \right)_c \right] \frac{RT_x^c}{V_x^c} \tag{7.42}$$

and

$$F_x = 3.46 + g(\eta) \tag{7.43}$$

where $g(\eta)$ is a function of η/k and is positive. Hence,

$$\frac{T^c - T_x^c}{T_x^c} = x_i x_j \left[\frac{C_1}{T_x^c} \left(\frac{\partial T_x^c}{\partial x_i} \right)_c + \frac{C_2}{V_x^c} \left(\frac{\partial V_x^c}{\partial x_i} \right)_c \right]^2 \tag{7.44}$$

where $C_1 = 1.7/(F_x)^{1/2}$ and $C_2 = 0.22/(F_x)^{1/2}$. For $\eta = 0$, $g(\eta) = 0$, $C_1 = 0.91$, and $C_2 = 0.12$. However, for nonspherical molecules, $g(\eta)$ in (7.43) increases, and both the constants C_1 and C_2 decrease. The van der Waals equation of state yields $C_1 = 0.75$, and it is a good compromise value. The second term in (7.44) may be neglected in most cases. The same treatment leads to the relations

$$\frac{V_m^c - V_x^c}{V_x^c} = -2x_i x_j \left[\left(\frac{C_1}{T_x^c} \frac{\partial T_x^c}{\partial x_i} \right)^2 - \left(\frac{C_2}{V_x^c} \frac{\partial V_x^c}{\partial x_i} \right)^2 \right] ; \tag{7.45}$$

$$\frac{P^c - P_x^c}{P_x^c} = \left(\frac{\partial \ln P}{\partial \ln T} \right)_{V^c, x} \frac{(T^c - T_x^c)}{T_x^c} \tag{7.46}$$

The last derivative was expressed by Pitzer and his coworkers (1955) for pure fluids as

$$(\partial \ln P / \partial \ln T)_V = 5.808 + 4.93\omega; \quad (V = V^c) \tag{7.47}$$

A better value of the first coefficient, 5.83 for argon, was obtained by Hall and Eubank (1976). The acentric factor of mixtures depends on the value of η_{ij}/k, but for the purposes above it suffices to assume $\omega_m = (\omega_i + \omega_j)/2$ in (7.47). The derivative $(\partial \ln P / \partial \ln T)_v$ is constant only in the nearest vicinity of T^c, and Eq. (7.46) is valid only for small $(T^c - T_x^c)/T_x^c$. The terms $(T^c - T_x^c)$ and $(P^c - P_x^c)$ are always positive, whereas $(V_m^c - V_x^c)$ is usually negative because the second term in (7.45) is very small.

When T_x^c, P_x^c, and V_x^c are properly related to $\bar{u}_m$ and σ_m, the method yields accurate critical locus curves, particularly for $T^c(x)$ and $P^c(x)$. For mixtures of nonspherical molecules, $\bar{u}_{ii}/\bar{u}_{jj}$ may be treated as independent of T only in the critical region. Kreglewski and Kay (1969) estimated that for large molecules, $\bar{u}_{ii}/\bar{u}_{jj} \approx (T_i^c/T_j^c)(V_i^\varrho/V_j^\varrho)^{1/3}$. The values from this correlation are a linear function of the values of

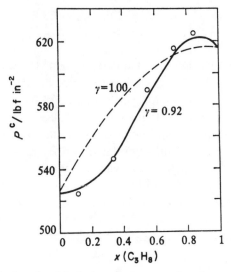

FIG. 7.24.   Calculated critical pressure curve for the propane + tri-*F*-acetonitrile. The experimental points were determined by Mousa et al. (1972). Reproduced with permission from *J. Chem. Thermod.* 4:301.

$\bar{u}/k$ obtained by Beret and Prausnitz (1975); this justification is demonstrated by Chen and Kreglewski (1977). By applying the square-well approximation for $\bar{u}_m$, the segment surface interactions, the combining rule (7.33) with $\gamma_{ij} = 1$, and the values given above for $\bar{u}_{ii}/\bar{u}_{jj}$, Kreglewski and Kay obtained $T^c(x)$ and $P^c(x)$ curves agreeing very well with the observed data for systems with London energies.

For systems with multipole interactions, $\gamma_{ij} \neq 1$. As an example, the $P^c(x)$ curve of the propane + trifluoro-acetonitrile determined by Mousa et al. (1972) is shown in Fig. 7.24. The value $\gamma_{ij} = 0.92$ was fitted to the $T^c(x)$ curve. The system is unusual in that the $P^c(x)$ curve exhibits two extrema. Another such system is the $C_3H_8 + SF_6$ system, examined by Clegg and Rowlinson (1955), for which $\gamma_{ij} = 0.90$, and the agreement with the observed $P^c(x)$ data is again satisfactory. In both these systems the molecular sizes of the components are very similar, so that $g_{11} \approx g_{22} \approx 1$ in (7.29). The large deviations of γ_{ij} from unity are due to the multipole interactions in the fluorinated components.

The results obtained for numerous systems show that Eq. (7.44) begins to fail for $\bar{u}_{jj}/\bar{u}_{ii} > 2$. The results obtained with the "classical" assumption that $\bar{u}_{jj}/\bar{u}_{ii} = T_j^c/T_i^c$ and that $\bar{u}_m$ is the average of total inter-actions (all $g_i = 1$) are reviewed by Hicks and Young (1975). In this case none of the existing combining rules for $\bar{u}_{ij}$ appears to be valid. Similar negative results were obtained by Young (1972) and by Dickinson and McLure (1974) for n-alkane + siloxane derivatives. Most of the deviations are due to improper weighting of the interactions.

The concept of pseudocritical constants is now rather obsolete. It was useful when the existing equations of state involved parameters that were intended to be clearly related to $\bar{u}/k$ and σ of the molecules but in fact are not, such as the van der Waals and Redlich-Kwong equations.

CHAPTER 8

Basic Experimental Studies of Mixtures at Low Pressures

Although most of this book is devoted to the theory and the calculation of phase equilibria in mixtures, the foundations of this branch of science are purely empirical. It seldom happens that the existence of a phenomenon is predicted as, for example, was the phenomenon of gas-gas equilibria by van der Waals and by Kammerlingh Onnes and Keesom. Nearly always a new phenomenon is observed first and interpreted later. Many thousands of systems have been examined, but most of them belong to the group of "complex mixtures" considered in the next chapter. Relatively few of them may be classified as simple systems, and their properties are sufficiently known to serve as a foundation for the development of a basic theory of mixtures. Many researchers made important contributions, but the following ones are well known, beginning with the 1890s: Kuenen, Zawidzki, Timmermans, Kay, Maass, Scatchard, Tsiklis, Reamer and Sage, Hildebrand, R. L. Scott, McGlashan, Staveley, Kohn, van Ness, Kobayashi, G. M. Schneider, G. C. Benson, Marsh, Streett, Grolier, and their coworkers. Their works are either pioneering and fundamental or distinguished by precision, the invention of improved methods, or a large volume of accurate experimental data.

Certain seemingly sound concepts may be disproved by a single well-conceived experiment. This chapter, then, will consider the experimental foundations: that is, the most important thermodynamic properties of certain selected systems which an acceptable theory should be able to explain and to correlate. Among these properties, the critical locus curves have already been discussed.

The essential part of some of the theories—namely, the prescription for averaging $\bar{u}_m$ based on the concept of segment interactions—has a shaky foundation. It may be valid only if the shell of interactions around a segment is really very thin and does not extend to other segments. We may guess only by comparisons with experimental data whether the concept of segment interactions is *useful*.

Tests of the validity of a given theory must be based on a sufficiently accurate equation of state. Errors in the liquid molar volumes of the *pure* components, calculated by means of the Gibbs conditions, are the most important indicators that the equation of state should be either improved or discarded.[†] The three-parameter principle of corresponding states (PCS) with the shape factors fulfills this criterion (Mollerup and Rowlinson, 1974; Teja and Rice, 1976) for substances not larger than hexane when methane is used as the reference substance. Among the equations of state considered in Chap. 6, (6.21), (6.24), and (6.26) conform to the criterion, the last of them being the most accurate equation. Eq. (6.57) is suitable over a limited range of temperatures, and the simplest equation, (6.18), is valid at high liquid densities, below the normal boiling point. The two are then useful for the evaluation of the excess functions of liquid mixtures or the solubility of gases in liquids at low pressures, when simplicity is desired.

Prior to comparing experimental data, it is proper to review briefly the methods of measurements of the thermodynamic functions of mixtures. The details of the equipment used are out of the scope of this book, so what follows is a review of reviews.

The enthalpy of mixing H^E and the excess heat capacity C_p^E are usually determined by calorimetric measurements. These are reviewed by McGlashan (1962) and by Marsh (1978). This review does not include the most recent calorimeter of Murray and Martin (1978). The excess volumes V^E of liquid mixtures are obtained from density measurements, and those of gaseous mixtures from compressibility factors. The measurements must be very precise because V^E/V_m is very small. The excess chemical potentials μ_i^E and G^E are usually calculated from liquid-vapor equilibrium isotherms $P(x)$ of mixtures at low pressures. The methods of the pressure measurements have repeatedly been reviewed, most recently by Marsh (1978) and by Malanowski (1982). The apparatus of Gibbs and van Ness (1972) is particularly convenient and precise. A very simple apparatus that is useful when two liquid phases are present was invented by Davison, Smith, and Chun (1967). Another precise apparatus was invented by Maher and Smith (1979) and Rogalski and Malanowski (1980).

[†] Regarding the densities of liquid natural gas, Haynes, Hiza, and McCarty (1977) state, "An error of 1% in density can result in an inequity of approximately \$40,000 per 125,000 m³ shipload of LNG. . . ." In most other cases the errors are less expensive, and the equation of state does not need to be so accurate.

The methods of measuring phase equilibria at high pressures are reviewed by Schneider (1978) and Young (1978), and the virial coefficients of mixtures by Knobler (1978). The gas-liquid critical curves may be obtained from *PVT* measurements; however, the most reliable methods in which $T^c(x)$ is observed directly and the system is properly stirred were worked out by Kay (1938), refined by Kay and Rambosek (1953), McMicking and Kay (1965), and Kay and Pak (1980). The methods and experimental problems in the measurements of the miscibility of liquids are discussed by R. L. Scott (1978). A method based on the determination of the dielectric parameters as a function of temperature is described by Riccardi and Sanesi (1966). It can be particularly useful at high pressures when the interface cannot be directly observed.

The conversion of observed $P(x)$ data to μ_i^E and G^E by means of Eq. (2.33) can be made only from low-pressure data, because the relations are exactly valid only at $P = 0$. The measurements of H^E and V^E are not limited to low pressures, but for any comparisons with G^E the results should be recalculated to $P = 0$—the difference between (2.21) and (2.26)—at least when the volatilities of the components are different.

At higher pressures, the $P(x)$, $P(x')$, $V_m(x)$, etc., curves must be calculated directly from the equilibrium conditions (2.11) and (2.12). By varying $\bar{u}_m$ and σ_m (depending on k_{ij}) until the calculated $P(x)$, etc., agree with the observed curves, we relate these curves *directly* to $\bar{u}_m$ and σ_m. The same calculations could be performed at low pressures, but it is easier to use G^E, H^E, and V^E at $P = 0$ as the *intermediate* functions of $\bar{u}_m$ and σ_m. The calculations are carried out as follows. If the characteristic constants of the pure components are not known, they are estimated by means of the general method outlined in Chap. 6. Next, with $Z_m = 0$, the volumes V_m of the system and V^E are obtained by an iteration at a given T, $\bar{u}_{ij}$, and $x_i = 0, 0.1, 0.2, \ldots$ from Eq. (6.26) or another equation of state. Given T, V_m, and x_i, the remaining excess functions are obtained directly from (2.24), (2.25), and (2.26). The values of $\bar{u}_{ij}$ and V_{ij}^o are either predicted from the combining rules or varied until the calculated G^E or H^E agree with the observed values.

In this chapter the comparisons are limited to systems with London energies only. The systems considered here are the key systems, and the properties (G^E, H^E, Henry's constant H_{ij}, V^E of the liquid and the vapor phases) are fundamental.

A. Systems of Small Molecules

The excess functions of six binary systems, predicted by using the equation of state (6.26), the square-well approximation with all $g_i = 1$ (mole fractions), $k_{ij} = 1$, and $\bar{u}_{ij}$ calculated from the combining rules (5.5) and (5.7), are compared with the observed values at $x = 0.5$ in Table 8.1. The approximate relation $\alpha_m = \sum_i x_i \alpha_i$ was used in this and the next sections. The characteristic constants of xenon and oxygen were estimated. In the case of the $N_2 + CH_4$ system, the value of $\eta_{11}/k = 3$ of nitrogen was ascribed entirely to its quadrupole moment, $\eta_{11}/k = \eta_{11}^Q/k$. Hence, $\eta_{12}/k = (\eta_{12}^L + \eta_{12}^Q) = 0.5$ K for this system because η_{22}^Q of methane and thus η_{12}^Q are zero. The agreement of the values of G^E and V^E with the observed values is very good. A deviation of a few $J \cdot mol^{-1}$ in G^E is hardly detectable on the liquid-vapor equilibrium isotherms.

The predicted values of the second virial coefficients (Eq. 6.36) of the Ar + Kr system agree very well with the values measured by Fender and Halsey (1962). Also, the critical locus curves of $N_2 + Ar$ and Ar + CH_4 based on the concept of pseudocritical constants are in excellent agreement with the data of Jones and Rowlinson (1963).

Thus it can be stated that the square-well (0) approximation and the rules (5.5) and (5.7) have been sufficiently tested, with positive results for the systems of small molecules except He, Ne, and H_2. Liquid-vapor equilibria of systems involving these elements appear at very low temperatures, where quantum corrections to the equation of state must be taken into consideration.

The excess functions of the same systems, predicted by using the van der Waals approximation, are only slightly less satisfactory. The two approximations become distinctly different when applied to systems of large molecules.

B. Systems of Large Molecules

The borderline between "small" and "large" molecules is not sharp. As the molecular size of one or of both components of a binary system increases, the errors in the predicted values of G^E may become as large as 3000 $J \cdot mol^{-1}$. This does not mean that the rules (7.31) to (7.33) fail, because the errors can be greatly reduced by operating with surface interactions and surface fractions and by adding A^{EC} to A_m^r. Even then the errors in G^E may exceed the acceptable error of, say,

TABLE 8.1. The excess functions of liquid systems of small molecules at $x = 0.5$.

System	Observed values				Predicted values[a]		
	T/K	G^E	H^E	V^E	G^E	H^E	V^E
Ar + Kr	103.94	82.5[c]	—	—	87	47	−0.19
	115.77	83.9[c]	—	−0.53[c]	93	26	−0.51
	115.77	82.4[d]	—	−0.464[d]	93	26	−0.51
Kr + Xe	161.36	103.0[d]	—	−0.459[d]	135	64	−0.51
	161.36	114.0[e]	—	−0.695[e]	135	64	−0.51
Ar + O$_2$	83.82	37.0[f]	60.0[f]	+0.14[f]	35	50	+0.12
Ar + CH$_4$	91.0	72.0[g]	103.0[h]	+0.18[h]	79	100	+0.15
	115.77	76.0[i]	—	—	79	69	−0.12
Kr + CH$_4$	115.77	29.0[j]	55.0[j]	−0.025[j]	22	22	−0.02
N$_2$ + CH$_4$	90.7	169.6[k]	—	—	166[b]	190[b]	−0.71[b]
	91.5	—	138.0[k]	—	166[b]	186[b]	−0.75[b]
	105.0	—	101.0[k]	−1.16[m]	169[b]	86[b]	−1.95[b]

NOTE: G^E and H^E are in $J \cdot mol^{-1}$; V^E in $cm^3 mol^{-1}$.

THEORY: Square-well approx.; all $g_i = 1$; Eqs. of state (6.26) to (6.29).

(a) Values calculated for $k_{ij} = 1$, combining rules (5.5) and (5.7).

(b) Values calculated with the assumption that η/k of N$_2$ is due to the quadrupole moment.

SOURCES: Reproduced with permission from *Adv. Chem. Ser.* 182:197; (c) Duncan, Davies, Byrne, and Staveley (1966); (d) Chui and Canfield (1971); (e) Calado and Staveley (1971); (f) Pool et al. (1962); (g) Sprow and Prausnitz (1966); (h) Lambert and Simon (1962); (i) Calado and Staveley (1972); (j) Calado and Staveley (1971); (k) McClure, Lewis, Miller, and Staveley (1976); (m) Massengill and Miller (1973).

$\pm$ 30 J $\cdot$ mol^{-1}. They may be attributed to the uncertainty of the values of g_i and η_{ii}/k of the pure components rather than to the failure of the combining rules. The values of $g_i^2 \sim (V_i^{oo})^{2/3}$ are crude approximations compared to the site fractions. The values of η_{ii}/k of molecules larger than hexane vary with ρ^*, as mentioned in Chap. 4, and those obtained from the acentric factors are often too large for a liquid at high densities $(T/T^c < 0.7)$.

The data for the important system, methane + ethane, at $x = 0.5$ are compared with the calculated values in Table 8.2. The errors in the predicted values are not negligible and γ_{ij} are adjusted here to enforce agreement with the observed value of G^E at one temperature (within $\pm$ 2 J $\cdot$ mol^{-1}). As before, the van der Waals and the square-well (0) approximation are distinguished by $\delta = 1$ and $\delta = 0$, respectively, in (7.29) and $k_{ij} = 1$. The values of H^E obtained for $\delta = 1$ are too small.

TABLE 8.2.   The excess functions of the methane + ethane system at $x = 0.5$.

T/K	G^E	H^E	V^E
Observed values			
91.5	114[a]	87[a]	—
108.2	—	—	−0.54[b]
112.0	122[a]	69[a]	—
Calculated: $\delta = 1$; all $g_i = 1$			
91.5	(112)	−10	−0.38
108.2	131	+22	−0.55
112.0	135	26	−0.60
Calculated: $\delta = 0$; all $g_i = 1$			
91.5	(113)	78	−0.38
108.2	119	79	−0.55
112.0	121	77	−0.61
Calculated: $\delta = 0$; $g_i^2/g_1^2 = (V_i^{oo}/V_1^{oo})^{2/3}$, A^{EC} included			
91.5	(114)	37	−0.37
108.2	126	59	−0.54
112.0	128	61	−0.59

NOTE: G^E and H^E are in J $\cdot$ mol^{-1}; V^E in cm^{3}mol^{-1}. Values calculated by the author (unpublished).
(a) Miller and Staveley (1976).
(b) Shana's and Canfield (1968).

The best results are obtained by using mole fractions instead of surface fractions, indicating that C_2H_6 belongs to "small" molecules. However, propane clearly belongs to "large" molecules. The values of G^E for the $CH_4 + C_3H_8$ system were determined by Calado et al. (1974) at 90.7 K (187 J · mol^{-1} at $x = 0.5$) and of H^E by Miller and Staveley (1976) at 91.5 K (154 J · mol^{-1}). In this case two parameters must be fitted to the experimental data, k_{ij} to V^E and γ_{ij} to G^E or H^E, and the results are better when surface fractions are used.

Marsh and coworkers measured with great precision the excess functions of numerous binary systems of globular molecules such as cycloalkanes (C_5 to C_8) + cycloalkanes or 2,3-dimethylbutane or octamethyl-cyclotetrasiloxane (OMCTS), a large molecule compared with the other ones. Marsh (1971) and Ewing and Marsh (1977) compared these data with those calculated by using several simple equations of state and various "prescriptions for mixing" the molecules with only one parameter k_{12}^u (in $\bar{u}_{12} = k_{12}^u(\bar{u}_{11}\bar{u}_{22})^{1/2}$) fitted to G^E at $x = 0.5$. Eq. (6.18) with the VDW approximation ($\delta = 1$) appears to represent best the asymmetry of the $G^E(x)$ curves. They assumed that the combining rule (5.11), that is,

$$(V_{ij}^o)^{1/3} = [(V_{ii}^o)^{1/3} + (V_{jj}^o)^{1/3}]/2 \tag{8.1}$$

holds for all these systems and obtained, for all the systems with OMCTS, values of V^E that were much too negative (except in the case of Flory's theory). Kreglewski et al. (1973, 1979) demonstrated that (8.1) yields wrong results. When the surface fractions and the surface interactions are used, γ_{ij} and k_{ij} (Eqs. 7.29 to 7.33) are close to unity for the foregoing systems.

Rodriguez and Patterson (1982) reviewed the existing large volume of data for n-alkane mixtures ($m_c > 5$) and have shown that the excess functions conform to the principle of corresponding states for polymers, including the Flory-Huggins relation for S^{EC}, which should also be included in the calculations for systems of globular molecules. In this case, however, it is not clear how to estimate r in $\phi_1 = N_1/(N_1 + rN_r)$; see Eq. 5.28. The "real" value of r may be smaller than the volume fractions suggest.

The temperature dependence of $\bar{u}$ was neglected in all the foregoing calculations. This is justified to some extent for long chains at high densities (see Eq. 4.46). Both the enthalpy of vaporization ΔH^v of pure n-$C_{16}H_{34}$ and the values of G^E of its mixtures with n-C_6H_{14} re-

TABLE 8.3. The values of η_{22}/k, $\bar{u}^o/k$, H^r_2 of n-hexadecane (for $\alpha = 1$, $C = 0.12$) and the values of G^E ($x = 0.5$) of the system with n-hexane at 333.15 K ($\gamma_{12} = 1$; $A^{EC}/RT = -0.1059$; $\delta = 0$).

$\dfrac{\eta/k}{K}$	$\dfrac{\bar{u}^o/k}{K}$	H^r/RT	$\dfrac{G^E}{J \cdot mol^{-1}}$
305 [a]	471	−13.5 [b]	—
200	560	−12.9 [b]	−12
150	611	−12.5 [b]	−75
0	832	−10.3 [b]	−367

(a) From Eq. (6.48).
(b) Of the liquid at $T/T^c = 0.6$.

quire a value of η_{22}/k of $C_{16}H_{34}$ less than 305 K, obtained from ω. The trials are shown in Table 8.3. For each value of η_{22}/k, a new value of $\bar{u}^o/k$ is needed to obtain correct liquid molar volumes using the Gibbs conditions (2.11) and (2.13). Each set gives a value of H^r_2 of pure $C_{16}H_{34}$ and a value of G^E ($x = 0.5$) for the $C_6 + C_{16}$ system, obtained by using Eqs. (7.29) to (7.33) with $\delta = 0$, $\gamma_{12} = k_{12} = 1$, $\delta_{12}/k = 0$, and η_{12}/k from the arithmetic mean rule. It appears that for $\eta_{22}/k \approx$ 150 K, both G^E and H^r_2 are very close to the observed values. The results are then internally consistent for this η_{22}/k.

From a practical point of view, the most important among the excess functions is G^E, directly related to the vapor pressure of a mixture and to the miscibility of the components. The vapor-liquid equilibria, particularly at elevated pressures, are described for engineering purposes by the *vaporization equilibrium ratios* defined by

$$K_i = x'_i/x_i; \quad K_j = x'_j/x_j; \text{ etc.,}$$

at a given pressure P. They vary with P (or T if taken under isothermal conditions) and are measures of the volatility of the components. The *relative volatility*, defined as

$$\alpha_{ij} = K_i/K_j = x'_i x_j / x'_j x_i \tag{8.2}$$

is the key quantity used to estimate the possible efficiency of a distillation. Their practical applications are outlined in the textbook by Chao and Greenkorn (1975).

The equilibrium ratios are insensitive to the accuracy of an equation of state, and most of them are suitable for evaluation of K_i. The K_i values give obviously less information about a phase diagram than x_i' and x_i separately. The K_i values have been often measured directly instead of x_i' and x_i (or the latter were not reported) and cannot be used to obtain a diagram, such as Fig. 7.7, where all the fine details—including the miscibility gap—can be viewed directly. The K_i values of a ternary system must be represented on a set of complex graphs, as shown by Wichterle and Kobayashi (1972). It seems that the presentation of a ternary system on a Gibbs concentration triangle (Fig. 10.2) would allow a more direct glimpse into the behavior of the system, and the K_i values could be easily determined graphically.

C. Solubility of Gases in Liquids: Very Dilute Solutions

The solubility of gases in liquids is not a separate problem, but it is an interesting case of phase equilibrium if the concentration of the dissolved gas is very small. Various measures of gas solubility have been proposed, but the most adequate one is obviously the mole fraction x_i of the gas in the liquid phase. Let us consider the methane + n-hexane system shown in Fig. 7.7. At these temperatures and pressures not exceeding about 2,000 psia, the vapor phase is nearly pure methane, whereas the composition of the liquid phase varies strongly with the pressure. At low pressures, all the bubble-point curves tend to become straight lines, $P/x_i =$ const., where $x_i \rightarrow 0$. The slopes of the lines vary with the temperature. Above T_1^c of CH_4 the vapor phase contains a "super-critical gas," whereas below T_1^c it contains vapors only, but it is superfluous to distinguish a gas from a vapor.

The concept of the excess functions is not applicable to gas solubilities, because it requires that the saturated vapor pressures of *all* the components are low at the given T. Gas solubilities must be calculated directly from Gibbs equilibrium conditions where the gas or the vapor is simply the less dense phase.

The solubilities of gases at $P = 1.01325$ bar in relatively non-volatile liquids have been extensively investigated by Hildebrand, Prausnitz, Clever, Battino, and their coworkers and in other centers. They are reviewed by Hildebrand et al. (1970), Prausnitz (1969), Wilhelm and Battino (1973), Clever and Battino (1975), and Wilcock et al. (1978). The apparatus used here is rather simple, but great care must be taken when stirring the liquid to prevent oversaturation with the gas, particularly when x_i is small and the liquid has a high surface tension (in the nitrogen + water system, for example). The amount of the solvent, n_j, and the volume of the dissolved gas, V_i, are measured directly. By subtracting the saturated vapor pressure of the pure solvent from total P, the partial pressure of the gas, p_i, is obtained. The amount of the dissolved gas, n_i, is then calculated from V_i, p_i, and Z_i. The compressibility factor at pressures not exceeding about 5 bar can be evaluated from (1.2), and at higher pressures from an accurate equation of state such as (6.24) or (6.26). The evaluation of x_i is here evidently based on the assumption that the gas phase is an ideal but imperfect gas mixture. It is accurate only at low pressures. As the pressure and the solubility increase, the usual bubble-point determinations become more suitable.

The limiting case of solubility is related to the thermodynamic functions through the Henry's constant H_{ij}, defined as

$$H_{ij}(T, P) = \lim_{x_i \to 0} (f_i/x_i) \tag{8.3}$$

where f_i and x_i are the fugacity and the mole fraction of the gas in the liquid phase. By this definition, instead of the phenomenological definition, $\lim_{x_i \to 0} (P/x_i)$, H_{ij} is directly related to the chemical potential through (1.42):

$$\ln\left(\frac{H_{ij}}{P}\right) = \lim_{x_i \to 0} \frac{(\mu_i - \mu_i^{pg})}{RT} = \frac{\mu_i^{r\infty}}{RT} \tag{8.4}$$

where superscript ∞ indicates infinite dilution ($x_i \to 0$). From the Gibbs-Helmholtz relation,

$$T[\partial(\mu_i/RT)/\partial T]_{P,x} = -h_i/RT$$

where h_i is the partial molar enthalpy of component i. Since for the perfect gas mixture $h_i^{pg}/RT = 1$ (from Eq. 1.14), we obtain

$$T\left[\frac{\partial \ln(H_{ij}/P)}{\partial T}\right]_{P,x} = T\left[\frac{\partial(\mu_i^{r\infty}/RT)}{\partial T}\right]_{P,x} = -\frac{h_i^{r\infty}}{RT} \tag{8.5}$$

where $h_i^{r\infty}$ is the partial residual enthalpy at infinite dilution (also called "the heat of solution"). Similarly, the thermodynamic relations $(\partial \mu_i/\partial P)_{T,x} = v_i$ and $[\partial(\mu_i^{pg}/RT)/\partial P]_{T,x} = 1/P$ lead to the result

$$P\left[\frac{\partial \ln H_{ij}}{\partial P}\right]_{T,\,x} = P\left[\frac{\partial(\mu_i^{r\infty}/RT)}{\partial P}\right]_{T,\,x} + 1 = \frac{Pv_i^\infty}{RT} \qquad (8.6)$$

where v_i^∞ is the partial molar volume at infinite dilution. Eqs. (8.4), (8.5), and (8.6) are thermodynamic identities which clearly relate H_{ij}/P to the intermolecular energy parameters $\bar{u}_{ij}$ and σ_{ij} through A_m^r and $(\partial A_m^r/\partial x_j)_{x_k}$ in Eq. (1.38).[†]

The pressures chosen in the determination of H_{ij} must be such that x_i is less than about 0.03 (Kreglewski, 1965). Unfortunately, the values of H_{ij} were usually determined at $P = 1$ atm, at which x_i is too large for many systems, and the limiting law (8.3) is probably not fulfilled.

The Gibbs equilibrium conditions and Eq. (1.39) can be applied with *non-zero* values of x_i (observed). The parameters $\bar{u}_{ij}$ and σ_{ij} are either predicted or adjusted to obtain the observed value of x_i. Once they are known, $\mu_i^{r\infty}/RT$ can be obtained from (1.38) by putting $x_i = 0$. For a mixture of two gases (i, j) in the same solvent, $\mu_i^{r\infty}$ and $\mu_j^{r\infty}$ will differ from $\mu_i^{r\infty}$ for a single gas i. The solubility of a gas is strongly affected by the presence of other gases. For this reason, the solvent must be carefully de-aerated.

In the foregoing procedure, the calculations of gas solubilities are part of the construction of a complete phase diagram $0 \leq x_i \leq 1$. They are simplified when limited to $x_i = 0$, and (H_{ij}/P) is calculated directly from the equation of state. Relations for (H_{ij}/P) were derived by Pierotti (1963) and Snider and Herrington (1967) from the simple Eq. (6.18), and by Cyzewski and Prausnitz (1976) from Eq. (6.24), with $c = 1$. Henry's constants were also evaluated by Neff and McQuarrie (1973) from the Leonard-Barker-Henderson perturbation theory (Eq. 4.1) and by Boublik and Lu (1978) from a similar theory extended by Boublik (1974, 1975) to hard convex bodies.

[†]The gas solubilities x_i are sometimes related to certain thermodynamic functions ΔG° and ΔH° for the changes in G and H in the process at constant T and P:

"i"(gas, 1 atm) $\rightarrow$ "i"(hypothetical solution, $x_i = 1$).

The state on the right-hand side is that of a liquefied pure gas. It cannot be obtained from an equation of state, according to which a gas remains a gas (at the same T and P), so ΔG° cannot be related to the intermolecular energy parameters.

A more interesting function is perhaps $\Delta\mu_i^r = \mu_i^{r\infty} - G_i^r$, where G_i^r is the potential of the pure gas at the same T and P.

As compared with the experimental data, the results obtained from these theories are less satisfactory than those from (6.18). Boublik and Lu ascribed the success of Eq. (6.18) to a fortuitous cancellation of errors. However, the poor results are partly due to neglecting three-body interactions in the above theories. Goldman (1978) used the perturbation theory, extended to include three-body interactions by Lee, Henderson, and Barker (1975), to evaluate H_{ij} of systems of noble gases. The values of $\bar{u}_{ij}$ and σ_{ij} given in Table 5.1 and 5.2 were used. The values of H_{ij} were thus *predicted* from independently obtained parameters. The results obtained either by including or by neglecting three-body interactions are shown in Fig. 8.1. The effect of three-body interactions may be stronger at $x_i \rightarrow 0$ than at higher concentrations of component i. The effect is partly canceled when molecules i and j are randomly dispersed in similar numbers, whereas at $x_i \rightarrow 0$ each molecule i is surrounded by molecules j exclusively.

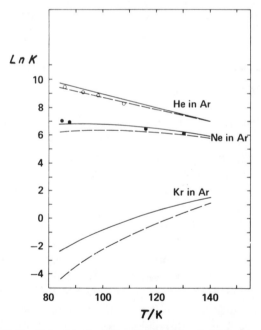

FIG. 8.1. $\ln H_{ij}$ (ln K in this figure) calculated by Goldman including three-body interactions (full lines) or neglecting them (dashed lines). The experimental points were determined by Street (1965) for Ne in Ar, and by Solen, Chueh, and Prausnitz (1970) for He in liquid Ar. Reproduced with permission from *J. Chem. Phys.* 69:3775.

TABLE 8.4.   Values of k_{ij}^u required to fit G^E ($x = 0.5$) and H_{ij} ($x_i = 0$) of systems of small molecules, according to Miller (1971).

	k_{ij}^u for G^E	k_{ij}^u for H_{ij}
Ar + CH$_4$	0.9757	0.98
Ar + Kr	0.9906	0.986
Ar + O$_2$	0.9884	0.99
Ar + N$_2$	0.9990	0.999
Ar + CO	0.9852	0.986
CO + CH$_4$	0.9817	0.998
CO + N$_2$	0.9898	0.99
N$_2$ + O$_2$	1.0040[a]	1.001[a]
Ne + Kr	0.79	0.79; 0.77
Ne + Ar	0.87	0.86
Ne + N$_2$	0.90	0.89

THEORY: Eq. (6.18) and the van der Waals approximation.
(a) These results are surprising, since N$_2$ and O$_2$ have quadrupole moments of the same sign.

The essential question, then, is this: are the values of $\bar{u}_{ij}$ and σ_{ij} fitting the properties at high concentrations of i—such as G^E, H^E, $T^c(x)$, etc.—also proper for $x_i \to 0$? Staveley (1970) and Miller (1971) compared the values of k_{ij}^u in the combining rule $\bar{u}_{ij} = k_{ij}^u (\bar{u}_{ii}\bar{u}_{jj})^{1/2}$, necessary to fit G^E at $x_i = 0.5$, with those required by the observed values of H_{ij} for the systems of small molecules. The equation of state used was (6.18). The more complete results of Miller are shown in Table 8.4, excluding the systems with H$_2$ and He, for which the values of G^E were determined at very low temperatures, and the quantum corrections to the hard-sphere equation of state were not known at that time.

These results show that H_{ij} of systems of small molecules may be estimated very well by using k_{ij}^u obtained from liquid-vapor equilibrium data at any concentration (but not vice versa, because the values of k_{ij}^u obtained from H_{ij} are less accurate). Similar results would probably be obtained by using a more accurate equation of state, such as (6.26). The effects of three-body (and multibody) interactions are here included because the D_{nm} constants (one constant in the case of Eq. 6.18) were obtained from the macroscopic properties of fluids. The agreement of the two sets of k_{ij}^u values, obtained by Miller, indicates that the three-body effects are about the same in the whole range of concentrations.

At the present time, there are no conclusive results for systems involving one or both components with large molecules. The combinatorial term A^{EC} tends to zero when $x_i \to 0$. If A^{EC} exists at other concentrations, the values of $\bar{u}_{ij}$, required to fit G^E or $T^c(x)$ including A^{EC}, should also be proper for evaluation of H_{ij}/P.

Another method of determining $\mu_i^{r\infty}$, based on the gas-liquid chromatography, was introduced by Everett and Stoddart and further developed by Cruickshank and coworkers. This field is reviewed by Locke (1976) and by Letcher (1978). Meyer and Baiocchi (1978) have found that the enthalpies of mixing at infinite dilution, calculated from chromatographic results for n-alkane mixtures, agree satisfactorily with the calorimetric values.

The most general method of determining x_i in very dilute solutions, not limited to gas solubilities, employs compounds labeled with radioactive ^{14}C or ^{35}S atoms. Magiera and Brostow (1971) studied extensively the liquid-vapor equilibria of binary systems of ^{14}C-labeled n-hexane with acetone, methyl-ethyl, methyl-propyl, and diethyl ketone at concentrations of hexane $x_i < 0.01$. Both bubble and dew points appear to be linear functions of x_i. The total (bubble) pressure P is

$$P/P_j = 1 + (b/P_j)x_i; \qquad (x_i \to 0) \tag{8.7}$$

where b is a constant at constant T, and P_j is the vapor pressure of pure j component. Along each of the isotherms, the ratio of the vapor to the liquid composition is constant:

$$x_i'/x_i = \text{const.}; \qquad (x_i \to 0)$$

By neglecting the variation of $(\mu_i^E)'$ in the vapor with x_i' (Eq. 2.35) in a dilute solution, they obtain for the liquid phase

$$[\partial(\mu_i^E/RT)/\partial x_i]_T = b/Px_j; \qquad (x_i \to 0) \tag{8.8}$$

The values of $\mu_i^{E\infty}/RT$ for the hexane + acetone system, obtained by Magiera and Brostow, are given in Table 8.5. The values in the third column were obtained by the extrapolation of μ_i^E obtained by Rall and Schaffer (1958) at high concentrations of i (hexane). The extrapolation was carried out by means of an empirical smoothing equation (the Redlich-Kister equation; see the next section). All the values are too large, as expected, because at high concentration μ_i^E varies approximately with x_j^2. Precise measurements of partial molar volumes and v_i^∞ were carried out by Handa and Benson (1982).

TABLE 8.5.   The excess chemical potentials $\mu_i^{E\infty}/RT$ of the hexane (i) + acetone system, according to Magiera and Brostow (1971).

T/K	Observed	Extrapolated
308.15	1.862	1.894
318.15	1.766	1.810
323.15	1.726	1.752
328.15	1.678	1.706

Traces of a sparingly soluble substance in a solution can be detected and the solubility determined from the pressure difference ΔP between the solution and the pure solvent as a function of the ratio of vapor to liquid volumes, V'/V. Ksiazczak and Buchowski (1980) derived the relation for the concentration of the solute:

$$x_{bi}/x_i = \Delta P_b/\Delta P = 1 + (\rho_j' P_i/\rho_j P_j x_i^s)(V'/V) \tag{8.9}$$

where x_{bi} and ΔP_b are the values of x_i and ΔP at the bubble point ($V' = 0$), ρ_j' and ρ_j are the saturated vapor and liquid densities of the pure solvent, P_i and P_j are the saturated vapor pressures, and x_i^s is the solubility of the solute at the given temperature. Thus, $1/\Delta P$ is a linear function of V'/V, and the slope yields x_i^s.

D. Smoothing of Liquid-Vapor Equilibrium Data

The results obtained at low pressures are usually expressed in terms of the excess functions: H^E and V^E are measured directly, whereas μ_i^E and G^E are calculated from the observed vapor pressures of mixtures. $G^E(x)$ is usually a nearly symmetrical function. In a crude approximation for a binary system, $G^E/x_1 x_2 \approx a_{12}$, where a_{12} is a function of the temperature only. The $V^E(x)$ and $H^E(x)$ curves are usually much less symmetrical than the schematic curves shown in Fig. 2.2.

The most frequently used method of determining $G^E(x)$ is due to Barker (1953). Here, $G^E/(x_1 x_2)$ is expressed by a power series of $(x_1 - x_2)^n$ ($n = 0, 1, 2, \ldots$), known as the Redlich-Kister equation (1948), and the coefficients a_n of the series are obtained by an iteration from total pressures (bubble points) and the mole fractions x_i in the liquid phase. This also yields the $\mu_1^E(x)$ and $\mu_2^E(x)$ curves and then the compositions of the vapor phase x_i' as explained in Chap. 2 in the discussion and application of Eqs. (2.32) and (2.33).

Wilson (1964) proposed the following, often used, equation:

$$G^E/RT = -x_1\ln(1 - ax_2) - x_2\ln(1 - bx_1) \tag{8.10}$$

where a and b are constants. This and other equations are analyzed critically by Monfort and Rojas (1978) and by Mollerup (1981). Extensive tables of liquid-vapor equilibrium data, smoothed by means of (8.10), have been prepared by Hirata, Ohe, and Nagahama (1975).

Marsh (1977) has recently shown that the existing relations for $G^E(x)$, which do not involve a logarithmic form, are special cases of the general equation

$$\frac{G^E}{RTx_1x_2} = \frac{\sum\limits_{m=0} a_m(1 - 2x_1)^m}{\sum\limits_{n=0} b_n(1 - 2x_1)^n} \tag{8.11}$$

where $b_o = 1$. He generalized the method of Barker accordingly. Once the coefficients of (8.10) or (8.11) are known, the values of x_i and x_i' can be recovered. They may differ slightly from the measured values, but they will be the best smoothed and thermodynamically consistent values.

Wieczorek and Stecki (1978) combined the van Laar (1910) equation for simple mixtures with the Mecke-Kempter equation (see Kehiaian and Treszczanowicz, 1968) for G^E due to association with one association constant K, G^E_{assoc},

$$G^E/RT = x_1x_2ab/(ax_1 + bx_2) + G^E_{\text{assoc}}/RT \tag{8.12}$$

where a, b, and K depend on T only. This equation appears to be very accurate and general because certain values of K can be ascribed (empirically) not only to mixtures containing alcohols or acids but to any system with complex interactions.

Brewster and McGlashan (1973) worked out a method for evaluating $G^E(x)$ from dew-point data. Kohler (1957) proposed a method of evaluating $G^E(x)$ in the case of limited miscibility in the system, and Stecki and Jackowski (1976) did so in the case when a mixture is liquid but the pure components are solids.

Since G^E is a function of the temperature, the coefficients of Eq. (8.10) or (8.11) vary accordingly for a given system. However, this does not mean that by differentiation, according to the Gibbs-Helmholtz relation, the proper relation for the $H^E(x)$ curve will be obtained. The $H^E(x)$ and $V^E(x)$ curves are usually smoothed by using separate sets of coefficients. A procedure for interrelating the $H^E(x)$ with the $G^E(x)$ curves was proposed by Liebermann and Fried (1972). The problem is again analyzed by Ochi and Lu (1978) and by Rogalski and Malanowski (1977).

According to the basic relation for G^E, (2.25), it depends also on ρ_i^* of the components, and these variables do not appear in the smoothing equations. The smoothing equations are designed to give the "best curves," and no physical meaning should be ascribed to the coefficients.

It was mentioned in Chap. 2 that the dew points are usually measured less accurately than the bubble points. Kreglewski (1971) stated that the dew-point curves calculated from the bubble-point data are more accurate than the directly measured values (at low pressures). Van Ness and coworkers (1973) concluded that such calculations are the most reliable tests of the thermodynamic consistency of the data at low pressures. They state, however, that none of the consistency tests is completely reliable. The various aspects of these problems were discussed by Abbott and van Ness (1977) and in a volume edited by Deyrieux and Kehiaian (1976).

The thermodynamic consistency test of equilibria at high pressures is a more complicated procedure. The problem was extensively studied by Prausnitz and his coworkers and is outlined by Prausnitz (1969). Christiansen and Fredenslund (1975) developed the *method of orthogonal collocation*. In this method the vapor compositions x_i' are calculated analogously to the tests at low pressures, from experimental values of P and x_i at a given T. It starts with the basic Eq. (1.35), and the required partial molar volumes and certain reference fugacities are obtained by the Prausnitz-Chueh (1968) correlation. The obtained values of x_i' are compared with the experimental values. The method was applied by Grauso, Fredenslund, and Mollerup (1977). For details about the consistency tests the reader is referred to the above publications.

Sets of experimental data for a given system are "reduced" to the few constants of an equation of state or a smoothing equation by the method of least squares or by methods based on the maximum likelihood principle of Fisher. The problems are considered by Neau and Peneloux (1981, 1982), who introduced the *successive reduction method*. This method allows the reduction of large sets of data by means of arrays of smaller subsets.

E. Compilations of Experimental Data

Besides the compilation of Hirata, Ohe, and Nagahama (1975), mentioned above, liquid-vapor equilibria at low pressures were collected by Hala et al. (1968) and are being systematically analyzed by Maczynski et al. (1976–82). A particularly valuable collection of both low- and high-pressure equilibrium data is due to Timmermans (1959). The critical locus data for binary systems were compiled by Hicks and Young (1975). Mixtures containing fluorocarbons (all equilibrium properties) are reviewed by Swinton (1978) in *Chemical Thermodynamics*. The same volume contains a bibliography of the thermodynamic properties of binary systems, edited by Hicks. A bibliography of the data for cryogenic fluid mixtures is due to Hiza, Kidnay, and Miller (1975). References can also be found in the *IUPAC Bulletin of Chemical Thermodynamics*, published since 1958.

A remarkable compilation of low-pressure liquid-vapor data, *International Data Series* (*Ser. A: Selected Data on Mixtures of Non-Reacting Species*) has been edited by Kehiaian and others and published by the Texas A&M Thermodynamics Research Center since 1973. It also contains data published for the first time and is thus both a compilation and a scientific journal.

A similar publication was started recently by Kertes (1979), covering the miscibility of gases, liquids, and solids. Liquid-liquid miscibility data were also compiled and critically analyzed by Francis (1963) and by Stein and Allen (1973).

CHAPTER 9

Complex Mixtures

Systems which exhibit *specific interactions* in addition to London energies will be called *complex mixtures*. The variety of specific interactions is staggering: there are dipole, quadrupole, and multipole interactions, electron donor + acceptor (DA) interactions (charge transfer complexes), hydrogen bridges ("bonds"), and other bridges. It is convenient to separate the DA interactions and to divide this chapter into two parts: (a) systems in which only one component exhibits specific interactions (or is capable of inducing them) and those remaining are *inert components* with London energies only (for example, the noble gases, alkanes, and cycloalkanes); (b) systems with at least two noninert components, including possible DA interactions.

At the present time we may state that the presence of specific interactions greatly affects the phase diagrams of mixtures. This is shown in Fig. 7.14, for example. The phase diagrams of pure fluids are significantly affected in the case of strong multipole interactions (as in fluorocarbons) and hydrogen bridges (as in water, ammonia, alcohols, alkanoic acids, and the lower aliphatic amines).

The hydrogen bridges manifest themselves in several striking ways. The experimental evidence is reviewed in the excellent monograph by Pimentel and McClellan (1960); the theories are outlined by S. H. Lin (1970) and by Kollman and Allen (1972). Pimentel and McClellan propose the following operational definition: an H-bond exists between a functional group A-H and an atom or a group of atoms B in the same or a different molecule when (a) there is evidence of bond formation, and (b) there is evidence that this new bond linking A-H and B specifically involves the hydrogen atom already bonded to A.

Hydrogen bridges may be *inter*molecular and *intra*molecular; in the second case they are formed between groups within a single molecule (chelation).

There is no convincing evidence that the nature of hydrogen bridges is other than electrostatic. Gubbins and Twu claim that their electrostatic theory of specific interactions holds also for water and ammonia. If the theory holds for water, it should also be suitable for other hydrogen-bonded liquids, because water is the least tractable substance.

Eq. (6.28) with the four parameters α, V^{oo}, $\bar{u}^o/k$, and $\eta = \eta^L + \eta^*$ holds in the cases of weak or moderate specific interactions. It does not hold, however, for water or methanol down to $T/T^c = 0.55$ (as it does for most substances) but only for $T/T^c > 0.77$ or 0.70, respectively (see Table 6.4). At lower temperatures, both the saturated liquid and vapor curves deviate from the experimental data. Clearly, in these cases the terms $A^{(2)}$ and $A^{(3)}$ from the Gubbins-Twu theory must be added to A^r calculated from Eq. (6.28).

Below, we limit our consideration to the systems for which the four-parameter Eq. (6.28) is approximately valid. Even then, our knowledge is so limited that the parameters $(\eta^L + \eta^*)$ and δ/k must always be obtained from some properties of the mixture under study: for example, G^E, H^E, or $T^c(x)$.

The success of the Gubbins-Twu theory has far-reaching consequences. It may mean that the *chemical theory of solutions* is superfluous. The theory was introduced by Dolezalek in 1908 to account mostly for the hydrogen "bonds" which are very strong compared to London interactions. It is outlined by Prausnitz (1969). Two types of "reactions" are distinguished here: *association*, or the formation of aggregates consisting of identical monomers (for example, in pure water), and *solvation*, or the formation of aggregates of two or more molecules of which at least two are not identical. Various models are conceivable to construct the equation for the "chemical" equilibrium and the equilibrium constant K_p. The dependence of K_p on the temperature is related to the standard enthalpy of the reaction $\Delta H^\ominus$. The most complete set of thermodynamic relations for virtually all types of association or solvation was derived by Kehiaian and his coworkers and applied in detail to the toluene + aniline system (Kehiaian and Sosnkowska-Kehiaian, 1966). Since the London energies of toluene and aniline are very similar, in the absence of association of aniline, the system would be nearly ideal. The system is treated here as an ideal associated mixture in which all the deviations from the ideal behavior are attributed to the association of aniline. The two parameters K_p and $\Delta H^\ominus$ were fitted to the observed values of G^E and H^E at one temperature and $x = 0.5$. The values of G^E, then calculated at other compositions and temperatures, agree well with the experimental data; and the derived hydrogen bond energies agree with the results obtained by infrared measurements,[†] but the results for H^E are rather poor.

As shown later, the electrostatic theory with two empirical parameters, $\eta = \eta^L + \eta^*$ and δ/k, leads to better results even for more complex systems. There is an essential difference in the application of the two theories. The association or solvation constant K is related to the *true* mole fractions of monomers, dimers, and so on. They obviously vary with the temperature and the density ρ^*, as does K. Evaluation of phase equilibria from Gibbs conditions, which involves the derivatives $\partial^n/\partial x_i^n$, $\partial^n/\partial V_m^n$ ($n = 1, 2$), becomes then very difficult. The electrostatic theory, in which x_i simply equals N_i/N, is free of such shortcomings.

In most cases, we are unable to predict thermodynamic properties of complex mixtures from those of the pure components alone. Among the possible approaches that attracted some attention are the *group contribution methods*. In these methods, the interaction energies, ε_{ij}, of each kind of atom or group of atoms in the molecule have a different value and give a constant contribution in any arrangement of the groups. This concept was introduced by Pierotti, Deal, and Derr (1959), Redlich, Derr, and Pierotti (1959), and Bondi and Simkin (1960) and was considered by Woycicki (1969). Further developments are reviewed by Kehiaian (1972, 1977), Baghdoyan and Fried (1975), Fredenslund, Jones, and Prausnitz (1975), Tsimering and Kertes (1975), Nitta et al. (1977), Wilson (1977), and Kehiaian et al. (1981).

The most developed method is UNIFAC, introduced by Prausnitz and coworkers (see Fredenslund et al., 1977). The excess potential μ_i^E of a component is assumed here to be the sum of two contributions: a combinatorial part, essentially due to the differences in size and shape of the molecules, and a residual part, due mostly to the interaction energy (not to be confused with the residual functions as defined in Chap. 1). The latter depends on the group interaction parameters a_{nm}, expressing an interaction between group n and group m. The choice and the size of the groups is a matter of experience. It may be as small as $C = 0$, whereas for the alcohols the optimal choice appears to be a group containing two carbon atoms; for example, CH_3OH and H_2O are

---

[†]In the interpretation of infrared spectra, as in the interpretation of thermodynamic data, it was *assumed* that there is a "chemical" reaction (association). It does not necessarily mean that "association" exists.

not subdivided. The values of a_{nm} were established from μ_i^E known for a large number of systems (liquid-vapor equilibria at low pressures). The UNIFAC method was modified by Nagata and Kawamura (1979) and by Nagata and Koyabu (1981). Among other results, they obtained correct liquid-liquid equilibrium curves in ternary systems, predicted from binary interaction parameters.

The group contribution methods differ in the way the parameters a_{nm} are related to μ_i^E. In the UNIFAC method, μ_i^E is expressed by Wilson's relation (8.10), with the constants a and b replaced by functions of the type $F_{nm} = \exp(-a_{nm}/T)$. Such "Boltzmann factors" appear when the mixture is assumed to be not random, but rather, highly ordered. On the other hand, Kehiaian related the group contributions to μ_i^E by means of Guggenheim equation (6.60) in which the nonrandom term is small. Fischer (1983) has shown on the basis of a perturbation theory, supported by Monte-Carlo simulation "experiments," that at high fluid densities the Guggenheim equation is essentially correct, whereas the substitution of the exponential functions F_{nm} into the Wilson equation yields completely wrong results. Even the systems with very large ratios of $\bar{u}_{22}/\bar{u}_{11}$, such as those shown in Figs. 7.16 to 7.21, are nearly completely random mixtures. When two liquid phases are present, each of them is a nearly random mixture except in the vicinity of the critical points.

Let us note that *the sign* of certain group contributions in molecule A may change, depending on the sign of the quadrupole moment in A relative to that in molecule B. For this reason, the group contribution methods should be limited to systems with only one quadrupolar molecule. Moreover, the total value of $\bar{u}/k$ depends on the temperature. Since it is not possible to determine separate values of η_{ij}/k and δ_{ij}/k for each group, the problem can be solved only by assuming that $\bar{u}/k$ is independent of T. This assumption limits the validity of a group contribution method to a narrow temperature range (near the normal boiling points). The methods would also be better if the evaluation of μ_i^E were based on the equation of state (6.61).

The systems in which all the components *except one* are inert components—such as the noble gases, alkanes, and cycloalkanes—have one common property: the deviations from an ideal mixture are always more positive than in a system of inert components with the same values of $\bar{u}_{ii}$, $\bar{u}_{jj}$, etc., of the pure components. This fact is easily explained by the electrostatic theory: η_{ij}^* and δ_{ij} arc zero because they are equal to zero for the inert components, so ε_{ij} is weak compared with the value that would result for the system with London energy only. An excellent example is the Xe + HCl system shown in Fig. 4.1.

Far more complicated are the systems with at least two components capable of specific interactions. Mixing such fluids may often but not always affect their electronic shells so that an additional attraction appears. The thermodynamic functions G^E and H^E of many such systems are smaller, $T^c(x)$ are higher, and $\beta(T, x)$ are more negative than expected either on a theoretical basis or by comparison with the functions of the systems formed by each of the components separately with inert solvents. Spectroscopic measurements, made for a great number of such systems, invariably point to the formation of *charge-transfer complexes* or *electron donor-acceptor* (DA) interactions between the two components. In their presence, the excess functions decrease and may even become negative if these interactions are strong enough.

According to Mulliken and Person (1969), the charge-transfer forces are short-range valence forces of quantum-mechanical origin. In this model, if the donor and the acceptor are even-electron systems, the ground state of the *weak* complex is described by a wave function:

$$\Psi_N(D, A) \approx a\underset{\text{(no-bond)}}{\Psi_o(D, A)} + b\underset{\text{(dative)}}{\Psi_1(D^+ - A^-)} \qquad (9.1)$$

where a and b are constants for a given system. The term $\Psi_o(D, A)$ contains the wave functions of the donor and the acceptor, each corrected for any polarization effects due to dispersion (London), repulsive *and* electrostatic forces. The term $\Psi_1(D^+ - A^-)$ is the *dative-bond* wave function, a relatively small correction term due to a partial transfer of an electron from the donor to the acceptor in the complex.[†]

The chemical theory of DA complexes has, therefore, a theoretical foundation. Spectroscopic equilibrium constants were determined for many systems, and some of the data are discussed by R. L. Scott (1971) and reviewed by Hildebrand, Prausnitz, and Scott (1970) and Drago (1974). The various kinds of electron donors and acceptors were classified by Mulliken and Person (1969). The most extensively investigated are those between halogens and aromatic hydrocarbons, initiated by Benesi and Hildebrand (1948, 1949), who discovered the

---

[†]The occasionally used terms "proton donor" and "proton acceptor" are unacceptable. There are complexes in which no protons are involved, but there is always a partial transfer of an electron. The "transfer" is actually a very small distortion of the electron orbital.

intense ultraviolet absorption in solutions of iodine in benzene and other aromatics.

In the complexes in which neither the acceptor nor the donor has a permanent dipole moment, the dipole moment of the complex (always present, as experiments indicate) is attributed to the *resonance interactions* between the two structures expressed by (9.1). Hanna (1968), Hanna and Williams (1968), and Lippert, Hanna, and Trotter (1969) have shown that all *induced* electrostatic interactions make a contribution to the energy and to the spectra of the benzene + halogen complexes, particularly the dipole moment in the halogen, induced by the quadrupole moment of benzene (the two components separately have zero dipole moments).

Let us note that halogens have positive quadrupole moments, whereas benzene has a negative one (Table 3.3). The leading term in $A^{(3)}$ (4.36) is then negative, and the total A_m^r becomes more negative than in the absence of this term. When A_m^r is more negative than the sum of $A_i^r x_i$, G^E is negative. Twu and Gubbins (1978) have shown that their theory predicts very well the liquid-vapor equilibria of the CO_2 + C_2H_4 system. Here, only the permanent quadrupole moments were taken into account. According to Mulliken, such interactions are included in $\Psi_o(D, A)$. It is possible that the Gubbins-Twu theory will not suffice in the case of complexes stronger than the CO_2 + C_2H_4 "complexes." If the appearance of the dative function Ψ_1 manifests itself by induced dipole *and* quadrupole moments, the proper values of Q_1 and Q_2 in $A^{(3)}$ (mixed interactions) will be the permanent plus the induced values. Even if the constants a and b in (9.1) were known for a given system, a useful relation of Ψ_N to the Helmholtz energy is not known, so the chemical theory cannot be applied for our purposes.

The electrostatic energy accompanying the formation of charge-transfer complexes is accurately related to A_m^r. Accordingly, the name "complex" will not be used further, although their existence is not denied.[†] We shall assume that the mole fractions obtained by weighing the components are also the true mole fractions (nonreacting systems).

[†] Weak complexes (i, j) in a fluid last for very brief periods of time. Split by collisions with other molecules, they are formed again when i meets j. The probability of such an encounter is the greatest when $x_i = x_j$. This is probably the reason why most weak complexes seem to contain the components in stoichiometric $1:1$ proportions. On the contrary, the melting curves may be the same as if a true compound had been formed (Ott et al., 1974, for example). *Clathrates*— and the most common of them, *hydrates*—should not be confused with complexes, as they are formed under proper mechanical conditions. Details about clathrates are outlined in the fundamental work by van der Waals and Platteeuw (1959).

Mulliken noted that the distinction between electron donors and acceptors is relative. For example, both benzene and the olefins are π donors. However, Vera and Prausnitz (1971) found that G^E of the 1-hexene + benzene system (at 303 K, $x = 0.5$) is smaller by about 140 J $\cdot$ mol^{-1} than for the *n*-hexane + benzene system. Since hexane and 1-hexene have very similar values of $\bar{u}$, the difference of G^E must be ascribed to the DA interactions between benzene and 1-hexene. Apparently, one of them behaves here as an acceptor. Let us note that C_6H_6 and C_2H_4 (and probably also other olefins) have Q_i of the opposite sign.

Gaw and Swinton (1968) studied the vapor-liquid equilibrium of the $C_6H_6 + C_6F_6$ system and found that G^E changes sign with the composition x_i. This is the only known system that forms simultaneously a positive and a negative azeotrope at the same temperature, as shown in Fig. 9.1. The excess enthalpies of this system were studied by Fenby, McLure, and Scott (1966, 1967) and Andrews and coworkers (1970), and the gas-liquid critical temperatures by Powell, Swinton, and Young (1970).

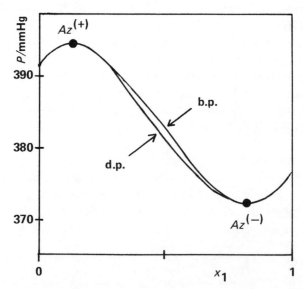

FIG. 9.1.   The dew- and bubble-point curves of the hexafluorobenzene (1) + benzene (2) system at 313.15 K, obtained by Gaw and Swinton (1968). The points Az$^{(+)}$ and Az$^{(-)}$ are, respectively, the positive and negative azeotropic points.

The excess enthalpies of C_6F_6 with aromatic or aliphatic amines were examined by Armitage and Morcom (1969) and by Mattingley, Handa, and Fenby (1975). The results obtained point to the existence of DA interactions in all these systems. The results noted above point to the following signs of Q_i of the substances: since that of C_6H_6 is negative, that of C_6F_6 should be positive; in turn, that of the amines should be negative. Therefore, the amines may not be expected to interact specifically with benzene. However, as estimated by Kreglewski and Wilhoit (1975) from the experimental data for G^E and H^E, they do interact *weakly* with benzene. This phenomenon, seemingly contradicting the electrostatic theory, has been noted for a few other systems. The phenomenon, as well as the asymmetry of the G^E and H^E curves often observed in systems with DA interactions, may eventually be explained as due to the induced quadrupole moment in the binary system. If its sign is the same as that of, say, Q_1 and opposite to that of Q_2, the interactions will be stronger at $0.5 < x_2 < 1$ but weaker at $0 < x_2 < 0.5$ than in the absence of the induced moment.

New techniques for studying specific interactions may yield more information about them. Among others, promising studies of the so-called hyper-polarizabilities were recently initiated by Levine and Bethea (1976).

Perfluoro-alkanes C_nF_m are only partially miscible with most organic liquids at low temperatures. They appear to be unable to interact specifically with any electron-donors or acceptors. In contrast, CCl_4 interacts with electron-donors, for example with olefins. The values of $H^E/J \cdot mol^{-1}$ at $x = 0.5$ and 298 K obtained by Woycicki (1975) for CCl_4 + hexenes are:

| | CCl_4 + | | | |
	1-*n*-hexene	*cis*-2-hexene	*trans*-2-hexene	*trans*-3-hexene
obs.	+91.	−33.	+41.	+3.
DA	(−220.)	(−275.)	(−245.)	(−250.)

where the values in parentheses are the DA contributions to H^E estimated by comparing the H^E of the above components with *n*-hexane.

If one or two of the fluorine atoms in a perfluoro-alkane are replaced by hydrogen or oxygen atoms, the compound is a moderately strong electron acceptor. Andersen et al. (1962) found that, while n-C_7F_{16} does not interact with acetone and the two are only partially miscible, 1-*n*-H.C_7F_{15} forms hydrogen bridges with acetone, $\rangle C{-}H^{(+)}$

. . . $^{(-)}O{=}C{<}$, where $(+)$ and $(-)$ indicate, respectively, a deficiency and an excess of the negative charge.

Acetic anhydride $(CH_3CO)_2O$ exhibits at most very weak interactions with acetone. However, Kreglewski (1963) and Stankiewicz, Kreglewski, and Woycicki (1963) found that the functions G^E, H^E, and V^E of the $(CF_3CO)_2O + (CH_3)_2CO$ system are all negative, indicating that the perfluoro-anhydride is an acceptor of a moderate strength. The behavior was attributed to a shift of electrons away from the terminal O atoms by the action of the F atoms, thus enabling the anhydride to interact with acetone through an *oxygen bridge* $>C{=}O^{(+)}$ . . . $^{(-)}O{=}C{<}$.

The amine $(C_4H_9)_3N$ is a weak donor. The effect of fluorine atoms on nitrogen is much weaker than on oxygen in a $>C{=}O$ group. The fluorinated derivative, $(C_4F_9)_3N$, is a rather inert compound (Kreglewski and Fajans, 1965).

Aliphatic ethers are still weaker donors. The values of G^E, H^E, C_p^E, and V^E, obtained by Murray and Martin (1978) for the mixtures of $(C_2H_5)_2O$ or $(C_3H_7)_2O$ with C_6F_6, point to at most very weak specific interactions between the components.

Sulfur dioxide is an acceptor and sometimes interacts strongly with donors. The values of G^E, determined by Lorimer, Smith, and Smith (1975) for the binary systems of SO_2 with aromatics, olefins, acetone, diethylaniline (all donors), and chloroform (an acceptor) point to moderate DA interactions in these systems. The interactions with olefins and aromatics were known earlier, and liquid SO_2 was used to extract them from petroleum fractions (the Edeleanu process). Interactions of SO_2 with $(C_2H_5)_2O$ were noted by Kreglewski and Jackowski (1966), and high-pressure equilibria and the critical locus $T^c(x)$, $P^c(x)$, and $V^c(x)$ curves were extensively investigated by Zawisza and Glowka (1967, 1970, 1971). Some of their results are shown in Figs. 9.2 and 9.3. A negative azeotrope is formed in the first system on all the isotherms, whereas it does not appear in the second system above about 50°C. This does not mean that the DA interactions are weaker in the second case. It is only a consequence of a larger difference in the vapor pressures of the pure components. Similarly, the differences in the critical temperatures of the pure components are too large for a maximum to appear in the first system, but it does appear in the second system.

Other known systems, stable at the critical point, with a maximum of $T^c(x)$ are the pyridine + acetic acid system (Swietoslawski

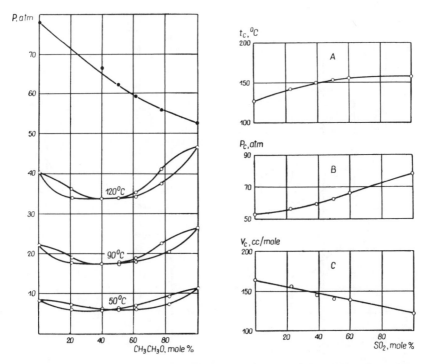

FIG. 9.2. The liquid-vapor equilibrium isotherms of the $(CH_3)_2O + SO_2$ system, determined by Zawisza and Glowka. A, B, and C are the critical locus curves $T^c(x)$, $P^c(x)$, and $V_m^c(x)$, respectively. The $P^c(x)$ curve is additionally shown (filled circles) in the first diagram. Reproduced with permission from *Bull. Acad. Polon. Sci.* 18:549, 555.

and Kreglewski, 1954), where it is due to a strong hydrogen bridge —O—H . . . N, and the cycloheptane + tetraethylsilane system (Hicks and Young, 1971), in which it may perhaps be due to a combinatorial contribution to A_m^r.

The behavior of carbon dioxide as either an electron acceptor or donor has been extensively investigated. High-pressure equilibria were measured for the $CO_2 + C_2H_4$ system by Ku and Dodge (1967), Mollerup (1975), and Fredenslund, Mollerup, and Hall (1976); Ohgaki and Katayama (1976), Tiffin et al. (1978), and Ng and Robinson (1978) made such measurements for the systems of CO_2 with aromatics. Kay and Kreglewski (1983) determined the critical locus $T^c(x)$ and $P^c(x)$ curves for the systems of CO_2 or SO_2 with benzene. The system $CO_2 + H_2O$ was studied extensively by Todheide and Franck (1963) and

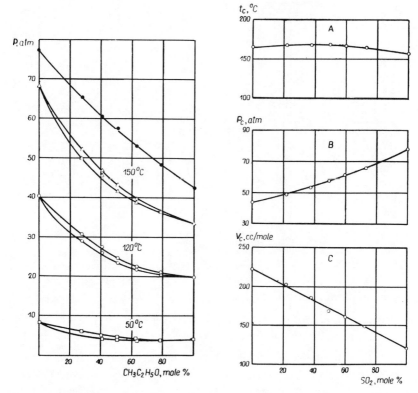

FIG. 9.3.   The liquid-vapor equilibrium isotherms of the $CH_3 \cdot O \cdot C_2H_5$ + SO_2 system, determined by Zawisza and Glowka. Conventions as in Fig. 9.2.

others. These data have not yet been analyzed. Glowka and Zawisza (1969) determined the critical locus curves and estimated the values of G^E of the CO_2 + perfluoro-acetone $(CF_3)_2CO$ system. The DA interactions between the two components are moderately strong. Presumably, a double oxygen bridge or "super-quadrupole" is formed in this system:

$$>C{=}O^{(+)} \ldots {}^{(-)}O{=}C{=}O^{(-)} \ldots {}^{(+)}O{=}C<$$

There are many known interesting systems with specific interactions for which most often H^E, less often G^E, and seldom $T^c(x)$ were determined. They have not been analyzed in recent years.

Attempts were made to extract from the observed values of the excess functions and the $T^c(x)$ curves the contribution due to DA interactions. For a given system at constant T and x_i, the observed functions may eventually be separated into two contributions:

$$G^E = G^{Eo} + G^{DA},$$

$$H^E = H^{Eo} + H^{DA},$$

$$T^c = T^{co} + T^{cDA}, \text{ etc.} \tag{9.2}$$

with G^{Eo}, etc., being the *hypothetical* excess functions in the absence of DA interactions, and G^{DA}, etc., being those due to the DA interactions. The values of G^{DA} and H^{DA} are always negative. The values of G^{Eo} and H^{Eo} may become negative when the differences in molecular sizes of the components are large; however, for mixtures of molecules of similar size, which are considered below, they are always positive.

In the case of the 1-hexene + benzene and the n-1-H.C$_7$F$_{15}$ + acetone systems, already discussed, the values of G^{Eo} and H^{Eo} were estimated directly by comparing with very similar noninteracting systems. In most cases it is difficult to find a proper inert solvent and to determine the reference excess functions of a noninteracting system. Hence, attempts were made to estimate G^{Eo}, etc., by applying the Scatchard-Hildebrand theory of regular solutions and the solubility parameters of the pure components (Harris and Prausnitz, 1969; Philippe and Clechet, 1973) or the quasi-chemical approximation, reviewed by Kehiaian (1972). Kreglewski and Wilhoit (1975) based the estimates of G^{Eo}, etc., on the hard-sphere equation of state with the simple attraction term (6.18), and $T^{co}(x)$ on the concept of the pseudocritical constants. They assumed that the observed values of η/k of the pure components are due entirely to the electrostatic interactions $\eta = \eta^*$ and $\eta^L = 0$. The importance of the $(\delta_{ij}/kT)^2$ term was not realized at that time. Nagata et al. (1981) expressed G^{Eo} by means of the nonrandom-two-liquid (NRTL) equation (see Renon and Prausnitz, 1968) and obtained very good results for systems involving acetonitrile, acetone, and chloroform. Each of these treatments yields different values of G^{Eo} for a given system.

The excess functions of complex mixtures can be evaluated without the separation of G^{Eo}, H^{Eo}, and so on. The functions calculated for the system chloroform + acetone by means of the equation of state (6.26), the relations (7.29) to (7.33)—$\delta = 0$; surface fractions—and adding μ_i^{EC} (Flory-Huggins) to μ_i are reported in Table 9.1. The characteristic constants estimated for chloroform are $V^{oo} = 49 \text{ cm}^3 \text{ mol}^{-1}$,

TABLE 9.1.   The excess functions of the system chloroform + acetone at $x = 0.5$.

	Observed values			Calculated values		
T/K	G^E	H^E	V^E	G^E	H^E	V^E
298	—	-1919[a]	-0.095[b]	-654	-2203	-0.56
313	-580[c], [d]	-1810[a]	—	(-581)	-1988	-0.59
323	—	-1740[a]	-0.20[e]	-538	-1861	-0.60
343	—	-1690[a]	—	-463	-1640	-0.65

SOURCES: (a) Morcom and Travers (1965); (b) Nigam, Mahl, and Singh (1972); (c) Zawidzki (1900); (d) Rock and Schroeder (1957); (e) Staveley, Tupman, and Hart (1955).

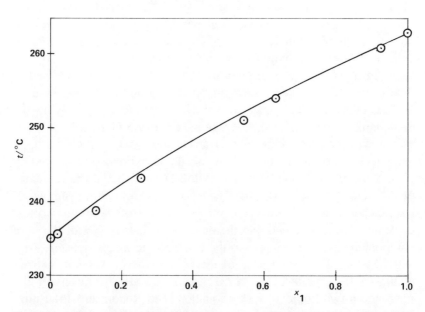

FIG. 9.4.   The gas-liquid critical temperatures of the chloroform + acetone system: experimental points determined by Swietoslawski and Kreglewski (1954) and the curve predicted from G^E data at low temperatures.

$\bar{u}^o/k = 504$ K, and $\eta/k = 75$ K (for $\alpha = 1$ and $C = 0.12$). Although both k_{ij} and δ_{ij}/k should be fitted to the experimental data for this system, the results are shown for $\gamma_{ij} = k_{ij} = 1$ and $\delta_{ij}/k = 77$ K adjusted to one value of G^E (given in brackets). The agreement is satisfactory, but the dependence of H^E on T is too strong. The calculated $T^c(x)$ curve

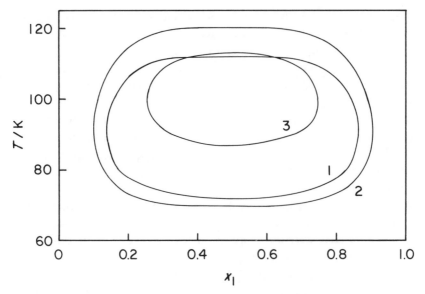

FIG. 9.5. Miscibility loops at saturation pressure obtained for the argon + argon system with $\gamma_{ij} = 0.7$ and

	k_{ij}	δ_{ij}/k
(1)	1.0	38.7
(2)	0.9	38.7
(3)	0.9	44.7

Reproduced with permission from *Fluid Phase Equil.* 9:205.

does not deviate much from the experimental data (Fig. 9.4). These results demonstrate the importance of δ_{ij}/k in the case of specific interactions between the components. For most systems with specific interactions, $k_{ij} \neq 1$ and must be adjusted to fit V^E.

Finally, let us consider the limited miscibility in liquid mixtures. The LCEP and UCEP in the gas-liquid critical region of the more volatile component (Fig. 7.9) may appear in systems with London energies of interaction. However, the appearance of a miscibility gap at temperatures well below the gas-liquid $T^c(x)$ is possible when the value of $\bar{u}_{ij}$ is lower than that due to the inequality of London energies of the pure components. Calculations based on an accurate equation of state have been carried out for model systems only. Twu, Gubbins, and Gray (1976) calculated the effect of pressure on T^{ucs} of the system argon + krypton as a function of the dipole or the quadrupole moments, imposed in the system. The dependence is relatively weak because the effect of pressure on the critical solution temperature depends mostly on $V_{ij}^o - k_{ij}$ in Eq. (7.31). Kreglewski and Hall (1982) calculated by means of the equation of state (6.26) the miscibility gaps in the argon + argon system, with imposed specific interactions simulated by variables γ_{ij} and δ_{ij}/k in Eq. (7.32). When $\gamma_{ij} < 1$ but $\delta_{ij}/k = 0$, a UCSP only may appear; $(\delta_{ij}/kT)^2$ increases the value of A_m^r and tends to homogenize the system at lower temperatures. An LCSP and a closed miscibility gap may appear only if $\delta_{ij}/k > 0$, according to this model. The loops shown in Fig. 9.5 were calculated for an unusually low value of $\gamma_{ij} = 0.7$ (to shift the loops to higher temperatures where the equation of state is more accurate). According to the electrostatic theory, the $(kT)^{-2}$ term increases the value of A_m^r when the quadrupole moments have opposite signs; the LCSP may appear for the same reason. For the argon + argon model, $\eta_{ii}/k = \eta_{ij}/k = 0$. It should be underlined that varying η_{ij}/k does not suffice to produce an LCSP.

CHAPTER 10

Some of the Problems in Separation of Mixtures

One of the commonly encountered problems in the separation of mixtures into pure components by distillation is azeotropy. From a theoretical point of view, there is nothing unusual about azeotropy. For the given values of $\bar{u}_{ii}$, $\bar{u}_{jj}$, and $\bar{u}_{ij}$, azeotropy may or may not appear, depending on the differences in the boiling temperatures (at constant P) or the vapor pressures (at constant T) of the components. However, azeotropy poses a difficult problem from the practical point of view because at the azeotropic point the compositions of the liquid $\bar{x}_i$ and the vapor phase $\bar{x}_i'$ are identical: $\bar{x}_i = \bar{x}_i'$. The azeotropic mixture cannot be separated by distillation, because the efficiency of the process depends on the difference between x_i and x_i'.

The azeotropic composition may be shifted in some cases in a favorable direction, to increase the yield of one of the components, by adding a properly chosen component to the mixture (extractive distillation) or by changing the pressure in the system. Most often, however, the distillation process must be combined with the extraction of the "bad" component in advance.

The situation becomes even worse when a so-called *tangent azeotrope* is formed in the system. The tangent azeotrope is defined as the azeotrope that has the composition and the boiling temperature of one of the pure components. Its adverse effect on the distillation process was first recognized by Swietoslawski (1963) and his coworkers, and the reasons were explained by Malesinski (1965). The n-hexane + benzene system, examined by Li, Lu, and Chen (1973) and shown in Fig. 10.1, is a beautiful example. The azeotropic point is very near $x_1 = 1$ of hexane over a range of temperatures, and the compositions of the liquid and vapor phases are hardly distinguishable in the range of x_1 from about 0.8 to 1. This is a severe case in which, evidently, a change of the pressure in the distillation process would not help. A distillation of a sample containing, say, 10 percent benzene would give a false impression that it is pure hexane. When benzene is mixed with isomeric heptanes or octanes with boiling points higher than that of

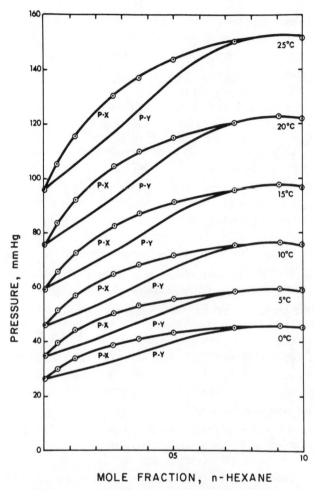

FIG. 10.1.   Liquid-vapor equilibrium isotherms of the *n*-hexane + benzene system, determined by Li, Lu, and Chen (1973). Reproduced with permission from *J. Chem. Eng. Data* 18:305.

n-hexane, tangent azeotropes may appear on the other side of the diagram, at $x_1 \to 0$. In this case, benzene could be wrongly believed to be pure. The existence of tangent azeotropes is probably the reason why samples of benzene of "spectroscopic purity" from different sources have different refractive indices. A reliable criterion of the purity of benzene is its freezing point.

As shown by Swietoslawski, tangent azeotropes are also formed by naphthalene with the phenols (phenol, cresols, and so on). In order to obtain naphthalene from coal tar in good yield, it is necessary to remove the phenols in advance by extraction (with aqueous sodium hydroxide solutions).

The possibility of a separation of two components from a mixture (and also of an azeotropic mixture treated as a component) depends on $(x_1' - x_1)$. Malesinski (1965) obtained the following relations for a binary system:

$$x_1' - x_1 = \left[\left(\frac{\partial x_1'}{\partial x_1} \right)_P^{Az} - 1 \right] (x_1 - \bar{x}_1)$$

$$+ \frac{1}{2} \left(\frac{\partial^2 x_1'}{\partial x_1^2} \right)_P^{Az} (x_1 - \bar{x}_1)^2 + \dots \tag{10.1}$$

where $(\partial x_1'/\partial x_1)_P^{Az}$ is the derivative at the azeotropic point $\bar{x}_1$, and the subscript P indicates that the problem is treated under isobaric conditions. If the vapor is a perfect gas, then

$$\left(\frac{\partial x_1'}{\partial x_1} \right)_P^{Az} = 1 + \bar{x}_1 \bar{x}_2 \frac{\partial}{\partial x_1} \left[\ln \frac{\gamma_1}{\gamma_2} \right] \tag{10.2}$$

where, as defined by (2.28), $\ln\gamma_i = \mu_i^E/RT$. The $G^E(x)$ curve is usually quite symmetrical and can be approximately expressed by the relation $G^E = a_{12}x_1x_2$, where a_{12} is a constant (depending on T). In this case $\ln\gamma_1 = a_{12}x_2^2/RT$ and $\ln\gamma_2 = a_{12}x_1^2/RT$, so that

$$\left(\frac{\partial x_1'}{\partial x_1} \right)_P^{Az} \approx 1 - 2\bar{x}_1 \bar{x}_2 \frac{a_{12}}{RT} \tag{10.3}$$

For an ordinary azeotrope, both terms in (10.1) contribute to $(x_1' - x_1)$, whereas for a tangent azeotrope $(\partial x_1'/\partial x_1)^{Az} = 1$, and the first term in (10.1) vanishes. Consequently, in the second case the compositions of the phases are nearly identical over a greater range of x_i than in the first case.

Another problem in the distillation of mixtures is due to the existence of ternary positive-negative, or the so-called *saddle azeotropes*. They may sometimes be formed when two of the components 1 and 2 exhibit electron donor + acceptor interactions, strong enough to cause

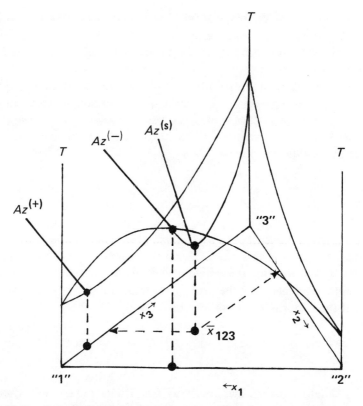

FIG. 10.2.   Schematic diagram of a system forming a saddle azeotrope $Az^{(s)}$. The curves shown are the boiling temperatures at a constant pressure. The composition of the saddle azeotrope is denoted by $\bar{x}_{123}$. The corresponding values of $\bar{x}_3$ and $\bar{x}_2$ are determined by the two arrows and dashed lines, parallel to the triangle sides (Gibbs triangle) and passing through $\bar{x}_{123}$ (clockwise convention).

the formation of a negative $Az^{(-)}_{12}$ azeotrope, whereas the two binary $(1 + 3)$ and $(2 + 3)$ systems exhibit positive deviations from an ideal system and *eventually* may form positive azeotropic mixtures, $Az^{(+)}_{13}$ or $Az^{(+)}_{23}$. The resulting boiling temperature surface is schematically shown in Fig. 10.2. The position of the saddle azeotrope depends on G^E_{12}, G^E_{13}, and G^E_{23} of the binary systems and on the boiling temperatures of the pure components. The separation of mixtures capable of forming saddle azeotropes is particularly difficult; it was extensively investigated by Swietoslawski and his students and analyzed by Malesinski.

Among the three main groups present in coal tar—namely, phenols, pyridine bases, and aromatic hydrocarbons—the components of the first two groups are capable of forming negative azeotropes. Aromatic hydrocarbons may exhibit only weak specific interactions with phenols or pyridine bases, and the deviations from an ideal mixture are positive in these systems. Consequently, the opportunity exists to form numerous saddle azeotropes in coal tar. Again, it is best to remove phenols in advance by an extraction.

The examples given above illustrate the main problems that could be solved with the aid of the theories of mixtures. Other problems are purely engineering problems and beyond the scope of this book. They are outlined in the textbooks by Holland (1963) and Smith (1963), and in the four volumes edited by Kobe and McKetta (1958–61). Tables of azeotropic mixtures were compiled by Horsley (1973).

Another problem is related to the number of components. Although the Gibbs equilibrium conditions can in principle be solved for a multicomponent system, it is impractical to go beyond a ternary system. One possible solution is to group all the components that, mixed among themselves, form ideal or nearly ideal mixtures into one *pseudocomponent*. For example, closely boiling isomeric alkanes will form nearly ideal systems; that is, the extremum values of G^E are not larger than about ± 30 J $\cdot$ mol^{-1} (Chen and Zwolinski, 1974). This treatment may be helpful in the calculation of distillation problems but may often be misleading in the treatment of mutual miscibility and, related to it, the separation of the components by extraction. The miscibility diagrams of ternary and pseudoternary systems are compared by Fleming and Vinatieri (1977). The author hopes that in the near future more research, based on the theories and the equations of state considered in this book, will be devoted to the problem of miscibility of fluids.

APPENDIX

Critical Temperatures T^c and Liquid Molar Volumes at $T/T^c = 0.6$

The critical temperatures T^c, given below, were taken from the compilations prepared by Kudchadker, Alani, and Zwolinski (1968), Mathews (1972), and Ambrose and Townsend (1975). Some of the values were slightly altered by the author.

Near the temperature $T/T^c = 0.6$, the vapor density is very small compared to the liquid density ρ_m and can be neglected. The Goldhammer relation is then

$$\rho_m = c[1 - (T/T^c)]^{0.3}$$

where c is a constant for the given substance. The constant can be evaluated from a single density datum at a temperature *close to $T/T^c = 0.6$*. The value of V_m at $T/T^c = 0.6$ is then calculated. These values, denoted by $V_m^{\ominus}$, are useful for estimating the close-packed volume, $V^{oo} \approx V_m^{\ominus}/1.7$, and $\bar{u}^o/k$ from the Gibbs equilibrium conditions. NOTE: The values in parentheses are either uncertain or correlated.

	T^c/K	$V_m^{\ominus}/\mathrm{cm^3 mol^{-1}}$
Oxygen	154.6	28.45
Hydrogen (normal)	33.25	28.3[†]
Water	647.3	19.025
Heavy water, D_2O	644.0	19.12
Helium-3	3.309	38.7[†]
Helium-4	5.189	28.4[†]
Neon	44.45	16.61[†]
Argon	150.86	29.14
Krypton	209.4	35.40
Xenon	289.75	45.59
Radon	377	56.9
Fluorine	144.3	34.37
Chlorine	417.2	46.28

[†]These values cannot be used to estimate V^{oo}.

Hydrogen fluoride	461	20.4
Hydrogen chloride	324.6	31.12
Hydrogen bromide	363.2	38.50
Hydrogen iodide	424.0	46.96
Sulfur dioxide	430.7	43.51
Hydrogen sulfide	373.2	36.03
Sulfur hexafluoride	318.7	(68.3)
Nitrogen	126.2	34.55
Nitric oxide, NO	180.3	22.06
Nitrous oxide, N_2O	309.6	(36.6)
Ammonia	405.6	25.16
Phosphine	324.8	40
Carbon monoxide	132.91	35.17
Carbon dioxide	304.20	33.88
Methane	190.58	38.46
Ethane	305.43	54.87
Propane	369.82	74.56
n-Butane	425.18	93.80
n-Pentane	469.5	113.2
n-Hexane	507.5	132.9
n-Heptane	540.3	152.6
n-Octane	568.8	172.3
n-Nonane	593.6	192.2
n-Decane	617.6	212.4
n-Undecane	638.7	232.5
n-Dodecane	657.7	252.9
n-Tridecane	676.2	273.6
n-Tetradecane	692.8	294.2
n-Pentadecane	709.2	315.2
n-Hexadecane	723	336
n-Heptadecane	736	357
n-Octadecane	(749)	378
n-Nonadecane	(760)	398
n-Eicosane	(770)	419
2-Methylpropane	408.13	94.96
2-Methylbutane	460.4	113.4
Neopentane	433.78	114.9
2-Methylpentane	497.45	132.9
3-Methylpentane	504.4	131.4
2,2-Dimethylbutane	488.78	132.8
2,3-Dimethylbutane	499.98	131.5
2,2,4-Trimethylpentane	543.96	172.1

Cyclopropane	398.25	59
Cyclobutane	460	78
Cyclopentane	511.61	95.85
Methylcyclopentane	532.73	116.3
Ethylcyclopentane	569.45	135.4
1,1-Dimethylcyclopentane	547	136
1,2-*cis*-Dimethylcyclopentane	565	134
1,2-*trans*-Dimethylcyclopentane	553	137
1,3-*cis*-Dimethylcyclopentane	551	137
1,3-*trans*-Dimethylcyclopentane	553	138
Cyclohexane	553.5	113.5
Methylcyclohexane	572.2	135.4
Ethylcyclohexane	609	154
Cycloheptane	604.3	130.7
Cyclooctane	647.2	147.5
cis-Decalin	702.3	172.9
trans-Decalin	687.1	177.2
Ethylene	282.35	49.26
Propene (Propylene)	364.85	68.17
1-Butene	419.53	87.18
1-Pentene	464.78	107.1
1-Hexene	504.03	126.9
1-Heptene	537.29	146.5
1-Octene	566.6	166.0
1-Nonene	593.3	187
1-Decene	617.1	207
cis-2-Butene	435.58	85.29
trans-2-Butene	428.63	87.08
2-Methylpropene	417.89	87.30
cis-2-Pentene	476	107
trans-2-Pentene	475	107
2-Methyl-1-butene	465	106
3-Methyl-1-butene	(450)	106
2-Methyl-2-butene	471	105
Propadiene	393	(62)
1,3-Butadiene	425	81
Acetylene	308.33	41.5
Propyne	402.38	58.6
1-Butyne	463.7	80.4
2-Butyne	488.6	78.28
1-Pentyne	493.4	98.58

Benzene	562.16	93.98
Toluene	591.80	114.0
Ethylbenzene	617.20	133.2
n-Propylbenzene	638.32	153.1
iso-Propylbenzene	631.1	152.5
o-Xylene	630.33	131.5
m-Xylene	617.05	133.2
p-Xylene	616.23	133.8
1-Methyl-2-ethylbenzene	651	150
1-Methyl-3-ethylbenzene	637	152
1-Methyl-4-ethylbenzene	640	153
1,2,3-Trimethylbenzene	664.47	148.7
1,2,4-Trimethylbenzene	649.17	151.2
1,3,5-Trimethylbenzene	637.25	152.2
Naphthalene	748.4	142.5
1-Methylnaphthalene	772	160
2-Methylnaphthalene	761	162
Dimethylether	400.1	62.23
Methyl-ethylether	437.9	81.3
Diethylether	466.74	101.7
Ethyl-*n*-propylether	500.2	121.7
Di-*iso*-propylether	500.3	142.8
Furan	490.2	72.64
2-Methylfuran	527	92.7
Tetrahydrofuran	540.2	84.4
p-Dioxan	587.2	(91.2)
Acetaldehyde	461	55.0
Methanol	512.64	41.25
Ethanol	513.92	59.27
1-Propanol	536.78	77.03
1-Butanol	563.05	95.79
1-Pentanol	588.15	114.6
2-Propanol	508.30	77.55
2-Methyl-1-propanol	547.78	96.04
2-Methyl-2-propanol	506.21	95.64
2-Butanol	536.05	94.42
Cyclopentanol	(591)	97.4
Cyclohexanol	(615)	114

Phenol	694.25	97.7
o-Cresol	697.55	115.3
m-Cresol	705.75	116.4
p-Cresol	704.55	116.2
Acetone	508.15	74.71
Methyl-ethylketone	536.78	93.02
Methyl-n-propylketone	561.08	113
Methyl-iso-propylketone	553.4	113
Diethylketone	561.46	111.6
Methyl formate	487.2	61.56
Ethyl formate	508.5	81.76
n-Propyl formate	538.0	101.2
n-Butyl formate	559	120
n-Pentyl formate	576	139
Methyl acetate	506.8	80.59
Ethyl acetate	523.2	100.8
n-Propyl acetate	549.4	120.7
n-Butyl acetate	579	141.0
iso-Butyl acetate	561	140
Methyl propionate	530.6	99.74
Ethyl propionate	546.0	120.2
n-Propyl propionate	578	140.6
Fluoromethane	317.8	38.24
Difluoromethane	335	41.5
Trifluoromethane	299.1	46.8
Fluoroethane	375.31	57.42
1,1-Difluoroethane	386.6	63.35
1,1,1-Trifluoroethane	346.2	79.38
1,1,1,2-Tetrafluoroethane	(324)	(80)
1,1,2,2-Tetrafluoroethane	(324)	(80)
Pentafluoroethane	(307)	(81)
1-H-Heptafluoropropane	377.6	103
1-H-Nonafluorobutane	(412)	132
1-H-Undecafluoropentane	444.0	159
1-H-Tridecafluorohexane	471.8	188
1-H-Pentadecafluoroheptane	495.8	215.5
Perfluoromethane	227.55	53.37
Perfluoroethane	293.2	81.5
Perfluoropropane	345.05	109
Perfluoro-n-butane	386.35	137.8

Perfluoro-*n*-pentane	420	164.7
Perfluoro-*n*-hexane	447.65	194.0
Perfluoro-*n*-heptane	474.8	221.1
Perfluoro-*n*-octane	502	249
Perfluoro-cyclobutane	388.37	115.2
Perfluoro-methylcyclohexane	486.75	194.0
Fluoroethylene	327.8	53.0
1,1-Difluoroethylene	303.2	57.0
Trifluoroethylene	305.0	60.2
Tetrafluoroethylene	306.4	63.7
Fluorobenzene	560.09	98.83
Pentafluorobenzene	531.95	113.6
Hexafluorobenzene	516.72	117.8
Hexafluoroacetone	357.25	96.65
Trifluoroacetic acid	491.28	76.72
Trifluoro-acetonitrile	311.11	69.0
Chloromethane	416.25	50.68
Dichloromethane	510	65.2
Chloroform	536.4	83.24
Carbon tetrachloride	556.4	101.6
Chloroethane	460.4	70.2
1,1-Dichloroethane	523	86.6
1,2-Dichloroethane	561	83.0
1,1,1-Trichloroethane	(580)	(100)
1-Chloropropane	503.2	89.24
2-Chloropropane	(493)	91.4
1-Chlorobutane	(542)	108
2-Chlorobutane	520.6	110
2-Chloro-2-methylpropane	(508)	112
Difluoro-chloromethane	369.2	(59)
Dichloro-fluoromethane	451.6	(71)
Trifluoro-chloromethane	302.05	67.07
Difluoro-dichloromethane	384.95	79.36
Fluoro-trichloromethane	471.15	90.8
1,1-Difluoro-1-chloroethane	410.2	(91)
Pentafluoro-chloroethane	353.2	(87)
Tetrafluoro-1,1-dichloroethane	418.6	(99)

Tetrafluoro-1,2-dichloroethane	418.8	(99)
Trifluoro-1,1,1-trichloroethane	(487)	(111)
Trifluoro-1,1,2-trichloroethane	487.2	111.0
Difluoro-1,1,1,2-tetrachloroethane	(551)	(129)
Difluoro-1,1,2,2-tetrachloroethane	551	129
Fluoro-pentachloroethane	(617)	(141)
Chlorobenzene	632.4	111.2
Bromomethane	464	55.3
Bromoethane	503.9	75.57
Bromobenzene	670.2	116.7
Iodomethane	528	64.25
Carbon disulfide	552	63.22
Dimethylsulfide	503.0	74.16
Methyl-ethylsulfide	533	93.7
Methyl-*n*-propylsulfide	(559)	113
Methyl-*iso*-propylsulfide	(550)	114
Diethylsulfide	557	113.5
2,3-Dithiabutane	(581)	94.0
3,4-Dithiahexane	649	135.4
Methanethiol	470.0	54.50
Ethanethiol	499	74.7
1-Propanethiol	(525)	93.2
1-Butanethiol	(551)	112
1-Pentanethiol	(572)	131
2-Propanethiol	514	95.6
2-Butanethiol	(541)	113
2-Methyl-1-propanethiol	(543)	113
2-Methyl-2-propanethiol	(524)	116
Triophene	579.4	84.31
Tetrahydrothiophene	632.0	96.3
Hydrogen cyanide	456.8	37.80
Acetonitrile	548.0	55.00
Propionitrile	564.4	74.95
n-Butyronitrile	582.2	93.81
Methylamine	430.0	44.08
Dimethylamine	437.6	65.20
Trimethylamine	433.3	94.95

Ethylamine	456.5	63.95
Diethylamine	496.6	104.6
Triethylamine	535.2	144.2
n-Propylamine	497.0	83.01
Di-*n*-propylamine	550	143
Tri-*n*-propylamine	577.5	200.5
n-Butylamine	524	101.0
Aniline	698.8	102.3
N-Methylaniline	701.7	121.8
N,*N*-Dimethylaniline	687.6	141.5
Pyrrole	639.8	75.8
1-Methylpyrrole	(667)	98.5
Pyridine	620.0	87.5
2-Methylpyridine	621	107
3-Methylpyridine	644.9	107
4-Methylpyridine	645.7	107
2,3-Dimethylpyridine	655.4	125
2,4-Dimethylpyridine	647	126.9
2,5-Dimethylpyridine	644.2	126
2,6-Dimethylpyridine	623.8	126
3,4-Dimethylpyridine	683.8	125
3,5-Dimethylpyridine	667.2	126
Piperidine	594.0	106
Nitromethane	588	(57.5)
Silane	269.7	43.2
Tetraethylsilane	603.7	202
Hexamethyl-disiloxane	518.6	217
Octamethyl-cyclo-tetrasiloxane	586.5	(332)
Silicon tetrafluoride	259.0	58.2
Boron trifluoride	260.9	43.0

References

Abbott, M. M., and H. C. van Ness. *Fluid Phase Equil*. 1977, 1:1.

Abe, A., and P. J. Flory. *J. Am. Chem. Soc*. 1964, 86:3563; 1965, 87:1838; 1966, 88:2887.

Adams, D. J., and A. J. Matheson. *J. Chem. Soc. Faraday Trans. II*. 1972, 68:1536.

Alder, B. J. *J. Chem. Phys*. 1964, 40:2742.

———; D. A. Young; and M. A. Mark. *J. Chem. Phys*. 1972, 56:3013.

Allen, G. D.; C. K. Chui; W. R. Madych; F. J. Narcowich; and P. W. Smith. *J. Approx. Th*. 1975, 14:302.

Altunin, V. V., and O. G. Gadetskii. *Teploehnerg. (U.S.S.R.)*. 1971, 18:81.

Ambrose, D., and R. Townsend. *Vapor-Liquid Critical Properties*. Teddington, Middlesex, U.K.: Nat. Phys. Lab., 1975.

Andersen, D.L.; R. A. Smith; D. B. Myers; S. K. Alley; A. G. Williamson; and R. L. Scott. *J. Phys. Chem*. 1962, 66:621.

Andersen, G. R., and J. C. Wheeler. *J. Chem. Phys*. 1979, 70:1326.

Andrews, A.; K. W. Morcom; W. A. Duncan; F. L. Swinton; and J. M. Pollock. *J. Chem. Ther*. 1970, 2:95.

Angus, S., and B. Armstrong. *International Thermodynamic Tables of the Fluid State: Argon*. London: Butterworth, 1971.

———; ———; and K. M. de Reuck. *International Thermodynamic Tables of the Fluid State: Carbon Dioxide*. London: Butterworth, 1976.

Armitage, D.A., and K. W. Morcom. *Trans. Faraday Soc*. 1969, 65:1.

Axilrod, B. H., and E. Teller. *J. Chem. Phys*. 1943, 11:299.

Baghdoyan, A., and V. Fried. *J. Chem. Ther*. 1975, 7:409, 895.

Baker, G. A., Jr. *Adv. Theor. Phys*. 1965, 1:1.

Barker, J. A. *Aust. J. Chem*. 1953, 6:207.

———. *Rev. Pure Appl. Chem*. 1955, 5:247.

——— and D. Henderson. *J. Chem. Educ*. 1968, 45:2.

——— and ———. *J. Chem. Phys*. 1967, 47:2856.

——— and ———. *Rev. Mod. Phys*. 1976, 48:587.

———; R. O. Watts; J. K. Lee; T. D. Schafer; and Y. T. Lee. *J. Chem. Phys*. 1974, 61:3081.

Barnes, A. N. M.; D. J. Turner; and L. E. Sutton. *Trans. Faraday Soc*. 1971, 67:2902, 2915.

Bartis, J. T. *J. Chem. Phys.* 1973, 59:5423.

Batsanov, S. S. *Russ. J. Inorg. Chem.* 1975, 20:2595.

Becker, F., and M. Kiefer. *Third International Conference on Chemical Thermodynamics*, vol. 2, p. 131. Baden, Austria, 1973.

Bedford, R. G., and R. D. Dunlap. *J. Am. Chem. Soc.* 1958, 80:282.

Bellemans, A. *Physica* 1973, 68:209.

———; V.Mathot; and M. Simon. *Adv. Chem. Phys.* 1967, 11:117.

———; ———; and P. Zuckerbrodt. *Acad. Roy. Belg. Bull. Cl. Sci.* 1956, 42:631, 643.

Bender, E. *The Calculation of Phase Equilibria from Thermal Equation of State*. Karlsruhe: Muller Verlag, 1973.

Benesi, A., and J. H. Hildebrand. *J. Am. Chem. Soc.* 1948, 70:2832; 1949, 71:2703.

Ben-Naim, A. In *Solutions and Solubilities*, ed. M. R. J. Dack, ch. 2. New York: Wiley, 1975.

Beran, J. A., and L. Kevan. *J. Phys. Chem.* 1969, 73:3860.

Beret, S., and J. M. Prausnitz. *A. I. Ch. E. J.* 1975, 21:1123; *Proc. 54th Convention of Gas Processors Assoc.* Tulsa, Okla., 1975.

Berne, B. J. In *Physical Chemistry*, vol. 8-B, ed. D. Henderson, ch. 9. New York: Academic Press, 1971.

——— and D. Forster. *Ann. Rev. Phys. Chem.* 1971, 22:563.

Bett, K. E.; J. S. Rowlinson; and G. Saville. *Thermodynamics for Chemical Engineers*. Cambridge, Mass.: MIT Press, 1975.

Bhattacharyya, S. N.; D. Patterson; and T. Somcynsky. *Physica* 1964, 30:1276.

Bienkowski, D. R., and K. C. Chao. *J. Chem. Phys.* 1975, 62:615.

Bierlein, J. A., and W. B. Kay. *Ind. Eng. Chem.* 1953, 45:618.

Block, T. E.; E. Dickinson; C. M. Knobler; V. N. Schumaker; and R. L. Scott. *J. Chem. Phys.* 1977, 66:3786.

Bondi, A. *Physical Properties of Molecular Crystals, Liquids and Glasses*. New York: Wiley, 1968.

——— and D. J. Simkin. *A. I. Ch. E. J.* 1960, 6:191.

Bothorel, P. *J. Colloid Interface Sci.* 1968, 27:529.

Boublik, T. *Collection Czech. Chem. Commun.* 1974, 39:2333.

———. *Fluid Phase Equil.* 1977, 1:37; 1979, 3:85.

———. *J. Chem. Phys.* 1975, 63:4084.

——— and B. C. Y. Lu. *J. Phys. Chem.* 1978, 82:2801.

Bradley, R. S. In *High Pressure Physics and Chemistry*, vol. 1, ch. 5. London: Academic Press, 1963.

Breedveld, G. J. F., and J. M. Prausnitz. *A. I. Ch. E. J.* 1973, 19:783.

Brewster, E. R., and M. L. McGlashan. *J. Chem. Soc. Faraday Trans. I.* 1973, 69:2046.

Brostow, W. *Chem. Phys. Let.* 1975, 35:387; 1977, 49:285.

———. *High Temp. Sci.* 1974, 6:190.

———. *Science of Materials.* New York: Wiley, 1979.

——— and Y. Sicotte. *Physica* 1975, 80-A:513.

Brown, I. *Ann. Rev. Phys. Chem.* 1965, 16:147.

Buckingham, A. D. In *Physical Chemistry*, vol. 4, ed. D. Henderson, ch. 8. New York: Academic Press, 1970.

———. *Q. Rev. Chem. Soc. London* 1959, 13:183.

——— and H. C. Longuet-Higgins. *Mol. Phys.* 1968, 14:63.

Byers Brown, W. *Phil. Trans. Soc. London Ser.* 1957, A-250:175, 221.

Cagniard de la Tour, C. *Ann. Chim.* (Paris) 1822, 21:127, 178; 1823, 22:410.

Calado, J. C. G.; G. A. Garcia; and L. A. K. Staveley. *J. Chem. Soc. Faraday Trans. I* 1974, 70:1445.

———; C. G. Gray; K. E. Gubbins; A. M. F. Palavia; V. A. M. Soares; L. A. K. Staveley; and C. H. Twu. *J. Chem. Soc. Faraday Trans. I* 1978, 74:893.

———; A. F. Kozdon; P. J. Morris; M. N. da Ponte; L. A. K. Staveley; and L. A. Woolf. *J. Chem. Soc. Faraday Trans. I* 1975, 71:1372.

——— and L. A. K. Staveley. *J. Chem. Soc. Faraday Trans.* 1972, 56:4718.

——— and ———. *Trans. Faraday Soc.* 1971, 67:289, 1261.

Calvert, R. L., and G. L. D. Ritchie. *J. Chem. Soc. Faraday Trans. II* 1980, 76:1249.

Carnahan, N. F., and D. E. Starling. *J. Chem. Phys.* 1969, 51:635.

Chandler, D. and L. R. Pratt. *J. Chem. Phys.* 1976, 65:2925.

Chang, J. I.; F. Hwu; and T. W. Leland. *Adv. Chem. Ser.* (in press).

Chao, J., and B. J. Zwolinski. *J. Phys. Chem. Ref. Data* 1973, 2:427.

Chao, K. C., and R. A. Greenkorn. *Thermodynamics of Fluids.* New York: Dekker, 1975.

Chapela, G. A., and J. S. Rowlinson. *J. Chem. Soc. Faraday Trans. I* 1974, 70:584

———; G. Saville; S. M. Thompson; and J. S. Rowlinson. *J. Chem. Soc. Faraday Trans. II* 1977, 73:1133.

Chen, C. H.; P. E. Siska; and Y. T. Lee. *J. Chem. Phys.* 1973, 59:601.

Chen, S. S., and A. Kreglewski. *Ber. Bunsenges. Phys. Chem.* 1977, 81:1048.

——— and B. J. Zwolinski. *J. Chem. Soc. Faraday Trans. II* 1974, 70:1133.

Christiansen, L. J. and A. Fredenslund. *A. I. Ch. E. J.* 1975, 21:49.

Chueh, P. L., and J. M. Prausnitz. *A. I. Ch. E. J.* 1967, 13:896.

Chui, C., and F. B. Canfield. *Trans. Faraday Soc.* 1971, 67:2933.

Clarke, H. A., and R. W. Missen. *J. Chem. Eng. Data* 1974, 19:343.

Clegg, H. P., and J. S. Rowlinson. *Trans. Faraday Soc.* 1955, 51:1333.

Clever, H. L., and R. Battino. In *Solutions and Solubilities*, ed. M. R. J. Dack, ch. 7. New York: Wiley, 1975.

Clifford, A. A.; E. Dickinson; P. Gray; and A. C. Scott. *J. Chem. Soc. Faraday Trans. I* 1975, 71:1953.

Cook, D., and J. S. Rowlinson. *Proc. Roy. Soc.* 1953, A-219:405.

Cox, J. D., and E. F. G. Herrington. *Trans. Faraday Soc.* 1956, 52:926.

Creek, J. L.; C. M. Knobler; and R. L. Scott. *J. Chem. Phys.* 1977, 67:366.

Croll, J. M., and R. L. Scott. J. Phys. Chem. 1958, 62:954.

Cyzewski, G. R., and J. M. Prausnitz. *Ind. Eng. Chem. F.* 1976, 15:304.

Dalgarno, A. *Adv. Chem. Phys.* 1967, 12:143.

Dantzler, E. M.; C. M. Knobler; and M. L. Windsor. *J. Phys. Chem.* 1968, 72:676.

Davenport, A. J., and J. S. Rowlinson. *Trans. Faraday Soc.* 1963, 59:78.

Davies, P. L., and C. A. Coulson. *Trans. Faraday Soc.* 1952, 48:777.

Davison, R. R.; W. H. Smith, Jr.; and K. W. Chun. *A. I. Ch. E. J.* 1967, 13:590.

de Boer, J. *Physica* 1949, 15:680.

Deiters, U. K. *Fluid Phase Equil.* 1982, 8:123; 1983, 12:193.

Delmas, G., and S. Turrell. *J. Chem. Soc. Faraday Trans. I* 1974, 71:572.

Denbigh, K. G. *Trans. Faraday Soc.* 1940, 36:936.

Desrosiers, W.; M. I. Guerrero; J. S. Rowlinson; and D. Stubley. *J. Chem. Soc. Faraday Trans. II* 1977, 73:1632.

de Swaan Arons, J., and G. A. M. Diepen. *J. Chem. Phys.* 1966, 44:2322.

Deyrieux, R., and H. V. Kehiaian, eds. *Reduction of Vapor-liquid Equilibrium Data*. Marseilles, France: U.E.R. Scientifique de Luminy, 1976.

Diaz Pena, M., and M. B. de Soto. *An. R. Soc. Esp. Fis. Quim., Ser. B. Quim.* 1965, 61:1163.

Dickinson, E., and J. A. McLure. *J. Chem. Soc. Faraday Trans. I* 1974, 70:2313.

Din, F. *Thermodynamic Functions of Gases*, vol. 1. London: Butterworth, 1956.

Dolezalek, F. *Z. Physik. Chem.* 1908, 64:727.

Donohue, M. D.; B. K. Kaul; and J. M. Prausnitz, *Proc. 56th Convention of Gas Processors Assoc.* Tulsa, Okla., 1977.

——— and J. M. Prausnitz. *Proc. 55th Convention of Gas Processors Assoc.* Tulsa, Okla., 1976.

Douslin, D. R., and R. H. Harrison. *J. Chem. Ther.* 1976, 8:301.

——— and ———. *J. Chem. Eng. Data.* 1977, 22:24.

———; ———; and R. T. Moore. *J. Phys. Chem.* 1967, 71:3477.

Drago, R. S. *J. Chem. Educ.* 1974, 51:300.

Duncan, A. G.; R. H. Davies; M. A. Byrne; and L. A. K. Staveley. *Nature* 1966, 209:1236.

Dunlap, R. D.; R. G. Bedford; J. C. Woodbrey; and S. D. Furrow. *J. Am. Chem. Soc.* 1959, 81:2927.

Dymond, J. H., and E. B. Smith. *The Virial Coefficients of Gases*. Oxford: Clarendon Press, 1969.

Eichinger, B. E., and P. J. Flory. *Trans. Faraday Soc.* 1968, 64:2035.

Einstein, A. *Ann. Phys.* 1910, 33:1275.

Eisenschitz, R., and F. London. *Z. Phys.* 1930, 60:491.

Ewing, M. B.; B. J. Levien; K. N. Marsh; and R. H. Stokes. *J. Chem. Ther.* 1970, 2:689.

——— and K. N. Marsh. *J. Chem. Ther.* 1977, 9:357, 863.

Eyring, H.; T. Ree; and N. Hirai. *Proc. Nat. Acad. Sci. U.S.A.* 1958, 44:683 (and subsequent papers on pure liquids).

Fannin, A. A., and C. M. Knobler. *Chem. Phys. Lett.* 1974, 25:92.

Fenby, D. V.; I. A. McLure; and R. L. Scott. *J. Phys. Chem.* 1966, 70:602; 1967, 71:4103.

Fender, B. E. F., and G. D. Halsey, Jr. *J. Chem. Phys.* 1962, 36:1881.

Fischer, J. *Fluid Phase Equil.* 1983, 10:1.

——— and S. Lago. *J. Chem. Phys.* 1983, 78:5750.

Fitts, D. D. *Ann. Rev. Phys. Chem.* 1966, 17:59.

Fleming, P. D., and J. E. Vinatieri. *J. Chem. Phys.* 1977, 66:3147.

Flory, P. J. *Discuss. Faraday Soc.* 1970, 49:7.

———. *J. Am. Chem. Soc.* 1965, 87:1833.

———. *J. Chem. Phys.* 1941, 9:660.

———. *Principles of Polymer Chemistry*. Ithaca, N.Y.: Cornell Univ. Press, 1953.

———; R. A. Orwoll; and A. Vrij. *J. Am. Chem. Soc.* 1964, 86:3507, 3515.

Flytzani-Stephanopoulos, M.; K. E. Gubbins; and C. G. Gray. *Mol. Phys.* 1975, 30:1481.

Fox, J. R., *J. Chem. Phys.* 1978, 69:2231.

Francis, A. W. *Liquid-Liquid Equilibriums*. New York: Interscience, 1963.

Fredenslund, A.; J. Gmehling; and P. Rasmussen. *Vapor-Liquid Equilibria Using UNIFAC*. Amsterdam: Elsevier, 1977.

———; R. L. Jones; and J. M. Prausnitz. *A. I. Ch. E. J.* 1975, 21:1086.

———; J. Mollerup; and K. R. Hall. *J. Chem. Eng. Data.* 1976, 21:301.

Freeman, K. S. C., and I. R. McDonald. *Mol. Phys.* 1973, 26:529.

Friend, J. A., J. A. Larkin; A. Maroudas; and M. L. McGlashan. *Nature* 1963, 198:683.

Frisch, H. L. *Adv. Chem. Phys.* 1964, 6:229.

Fulinski, A., and C. Jedrzejek. *Acta Phys. Pol.* 1974, A-46:591; *Physica* 1974, 78:173.

Garland, C. W., and K. Nishigaki. *J. Chem. Phys.* 1976, 65:5298.

Gaw, W. J., and R. L. Scott. *J. Chem. Ther.* 1971, 3:335.

———— and F. L. Swinton. *Trans. Faraday Soc.* 1968, 64:637, 2023.

Gibbons, R. M. *Mol. Phys.* 1969, 17:81.

Gibbs, J. W. *The Collected Works*, vol. 1. New Haven, Conn.: Yale Univ. Press, 1948.

————. *Trans. Conn. Acad.* 1876, 3:108.

Gibbs, R. E., and H. C. van Ness. *A. I. Ch. E. J.* 1972, 11:410.

Gillis, H. P.; D. C. Marvin; and H. Reiss. *J. Chem. Phys.* 1977, 66:214, 223.

Glowka, S. *Bull. Acad. Polon. Sci. Ser. Sci. Chim.* 1977, 25:239.

———— and A. C. Zawisza. *Bull. Acad. Polon. Sci. Ser. Sci. Chim.* 1969, 17:365.

Goldman, S. *J. Chem. Phys.* 1978, 69:3775.

Goodwin, R. D. *Adv. Cryog. Eng.* 1978, 23:611.

————. *J. Res. Nat. Bur. Stand.* 1975, 79-A:71.

————; H. M. Roder; and G. C. Straty. *Thermophysical Properties of Ethane.* NBS Tech. Note 684. Washington, D.C., 1976.

Gopal, R., and O. K. Rice. *J. Chem. Phys.* 1955, 12:2428.

Gosman, A. L.; R. D. McCarty; J. D. Hust. *NSRDS-NBS* 1969, 27.

Grauso, L.; A. Fredenslund; and J. Mollerup. *Fluid Phase Equil.* 1977, 1:13.

Gray, C. G.; K. E. Gubbins; and C. H. Twu. *J. Chem. Phys.* 1978, 69:182.

Greer, S. G., and R. Hocken. *J. Chem. Phys.* 1975, 63:5067.

Griffiths, R. B. *J. Chem. Phys.* 1965, 43:1958; 1974, 60:195.

———— and J. C. Wheeler. *Phys. Rev.* 1970, A-2:1047.

Gubbins, K. E., and C. H. Twu. *Chem. Eng. Sci.* 1978, 33:863.

———— and ————. In *Phase Equilibria and Fluid Properties in the Chemical Industry*, ch. 18. ACS Symp. Ser. 60. Washington, D.C., 1977.

Guerrero, M. I.; J. S. Rowlinson; and G. Morrison. *J. Chem. Soc. Faraday Trans. II* 1976, 72:1970.

Guggenheim, E. A. *J. Chem. Phys.* 1945, 13:253.

————. *Mixtures.* Oxford: Clarendon Press, 1952.

————. *Thermodynamics.* Amsterdam: North-Holland, 1967.

Gunning, A. J., and J. S. Rowlinson. *Chem. Eng. Sci.* 1973, 28:521.

Gurvich, L., et al. *Energy of Dissociation of Chemical Bonds, Ionization Potentials and Electron Affinity.* Moscow: NAUKA, 1974.

Haase, R. *Thermodynamik der Mischphasen.* Berlin: Springer Verlag, 1956.

Habgood, H. W., and W. G. Schneider. *Can. J. Chem.* 1954, 32:98, 164.

Haile, J. M. Ph.D. diss., Univ. of Florida, 1976.

Hala, E. *Ind. Eng. Chem. F.* 1975, 14:138.

————; J. Pick; and O. Vilim. *Vapor-Liquid Equilibrium.* New York: Pergamon Press, 1958.

————; I. Wichterle; J. Polak; and T. Boublik. *Vapor-Liquid Equilibrium Data at Normal Pressures*. London: Pergamon Press, 1968.

Hall, K. R. *J. Chem. Phys.* 1972, 57:2252.

———— and P. T. Eubank. *Ind. Eng. Chem. F.* 1976, 15:80, 323.

Handa, Y. P., and G. C. Benson. *Fluid Phase Equil.* 1982, 8:161, 181.

Hanna, M. W. *J. Am. Chem. Soc.* 1968, 90:285.

———— and D. E. Williams. *J. Am. Chem. Soc.* 1968, 90:5358.

Harris, H. G., and J. M. Prausnitz. *Ind. Eng. Chem. F.* 1969, 8:180.

Haynes, N. M.; M. J. Hiza; and R. D. McCarty. *Proc. 5th Conference on Liquid Natural Gas* (vol. 2, no. 11). Dusseldorf, Germany, 1977.

Heintz, A., and R. N. Lichtenthaler. *Ber. Bunsenges. Phys. Chem.* 1977, 81:921.

Hejmadi, A. V.; D. L. Katz; and J. E. Powers. *J. Chem. Ther.* 1971, 3:483.

Henderson, D., and J. A. Barker. *J. Chem. Phys.* 1968, 49:3377.

———— and S. G. Davison. In *Physical Chemistry*, vol. 2, ed. D. Henderson, ch. 2. New York: Academic Press, 1967.

———— and P. J. Leonard. In *Physical Chemistry*, vol. 8-B, ed. D. Henderson, ch. 7. New York: Academic Press, 1971.

Hicks, C. P., and C. L. Young. *Chem. Rev.* 1975, 75:119.

———— and ————. *Trans. Faraday Soc.* 1971, 67:1598.

Hildebrand, J. H.; J. M. Prausnitz; and R. L. Scott. *Regular and Related Solutions*. New York: Van Nostrand Reinhold, 1970.

———— and R. L. Scott. *Regular Solutions*. Englewood Cliffs, N.J.: Prentice-Hall, 1962.

———— and ————. *The Solubility of Non-Electrolytes*. New York: Reinhold, 1950.

Hill, T. L. *Introduction to Statistical Thermodynamics*. London: Addison-Wesley, 1960.

Hirata, M.; S. Ohe; and K. Nagahama. *Vapor-Liquid Equilibria*. Amsterdam: Elsevier, 1975.

Hirschfelder, J. O.; C. F. Curtiss; and R. B. Bird. *Molecular Theory of Gases and Liquids*. New York: Wiley, 1954.

Hissong, D., and W. B. Kay. *A. I. Ch. E. J.* 1970, 16:580.

Hiza, M. J.; A. J. Kidnay; and R. C. Miller. *Equilibrium Properties of Fluid Mixtures: A Bibliography*. New York: Plenum, 1975; vol. 2, 1982.

Hohenberg, P. C., and M. Barmatz. *Phys. Rev.* 1972, A-6:289.

Holland, C. D. *Multicomponent Distillation*. Englewood Cliffs, N.J.: Prentice-Hall, 1963.

Holleran, E. M. *J. Chem. Ther.* 1970, 2:779.

Horsley, L. H. *Azeotropic Data III*. Washington, D.C.: Amer. Chem. Soc., 1973.

Hougen, O. A.; K. M. Watson; and R. A. Ragatz. *Chemical Process Principles: Part II*. New York: McGraw-Hill, 1959.

Hsu, C. S.; L. R. Pratt; and D. Chandler. *J. Chem. Phys.* 1978, 68:4213.

Huff, J. A., and T. M. Reed III. *J. Chem. Eng. Data*. 1963, 8:306.

Huggins, M. L. *J. Chem. Phys.* 1941, 9:440.

————. *J. Phys. Chem.* 1939, 43:1083.

————. *J. Phys. Chem.* 1970, 74:371; 1971, 75:1255; *Macromolecules* 1971, 4:274; *Polymer* 1971, 12:389.

————. *Physical Chemistry of High Polymers*. New York: Wiley, 1958.

Hurle, R. L.; F. Jones; and C. L. Young. *J. Chem. Soc. Faraday Trans. II* 1977, 73:613.

Huron, M. J.; G. N. Dufour; and J. Vidal. *Fluid Phase Equil.* 1977–78, 1:247.

Jacobsen, R. T., and R. B. Stewart. *J. Phys. Chem. Ref. Data* 1973, 2:757.

Joffe, J., and D. Zudkevitch. *Chem. Eng. Prog. Symp. Ser. 81*. 1967, 63:43.

Johnston, K. P., and C. A. Eckert. *A. I. Ch. E. J.* 1981, 27:773.

————; D. H. Ziger; and C. A. Eckert. *Ind. Eng. Chem. Fund.* 1982, 21:191.

Jones, J. W., and J. S. Rowlinson. *Trans. Faraday Soc.* 1963, 59:1702.

Kac, M.; G. E. Uhlenbeck; and P. C. Hemmer. *J. Math. Phys.* 1963, 4:216, 229.

Kammerlingh Onnes, H., and W. H. Keesom. *Commun. Phys. Lab. Univ. Leiden* 1906, 96a–c; Nos. 15, 16.

Karnicky, J. F.; H. H. Reamer; and C. J. Pings. *J. Chem. Phys.* 1976, 64:4592.

Kaul, B. K., and J. M. Prausnitz. *Ind. Eng. Chem. F.* 1977, 16:335.

Kay, W. B. *Ind. Eng. Chem.* 1936, 28:1014; 1938, 30:459; 1940, 32:353; 1948, 40:1459.

———— and D. B. Brice. *Ind. Eng. Chem.* 1953, 45:615.

———— and A. Kreglewski. *Fluid Phase Equil.* 1983, 11:251.

———— and S. C. Pak. *J. Chem. Ther.* 1980, 12:673.

———— and G. M. Rambosek. *Ind. Eng. Chem.* 1953, 45:221.

Keesom, W. H. *Physik. Z.* 1921, 22:129.

Kehiaian, H. V. *Ber. Bunsenges. Phys. Chem.* 1977, 81:908.

————. In *Thermochemistry and Thermodynamics*, ed. H. A. Skinner, ch. 5. London: Butterworth, 1972.

————; J. P. Grolier; M. R. Kechavarz; and G. C. Benson. *Fluid Phase Equil.* 1980, 5:159.

————; ————; ————; ————; O. Kiyohara; and Y. P. Handa. *Fluid Phase Equil.* 1981, 7:95.

———— and K. Sosnkowska-Kehiaian. *Trans. Faraday Soc.* 1966, 62:838 (and the references given in this paper).

———— and A. Treszczanowicz. *Bull. Acad. Polon. Sci. Ser. Sci. Chim.* 1968, 16:445.

Kertes, A. S., ed. *Solubility Data Series*. London: Pergamon Press, 1979.

Kielich, S. *Chem. Phys. Lett.* 1971, 10:516.

————. *Physica* 1965, 31:444.

Kihara, T. *Intermolecular Forces*. New York: Wiley, 1976.

————. In *Physical Chemistry*, vol. 5, ed. H. Eyring. New York: Academic Press, 1970.

————. *Rev. Mod. Phys.* 1953, 25:831.

Kincaid, J. M.; G. Stell; and E. Goldmark. *J. Chem. Phys.* 1976, 65:2172.

King, M. B. *Phase Equilibrium in Mixtures*. Oxford: Pergamon Press, 1969.

Kirkwood, J. G.. *J. Chem. Phys.* 1935, 3:300.

———— and F. P. Buff. *J. Chem. Phys.* 1951, 19:774.

———— and Z. W. Salsburg. *Discuss. Faraday Soc.* 1953, 15:28.

Kirouac, S., and T. K. Bose. *J. Chem. Phys.* 1973, 59:3043.

Klein, H., and D. Woermann. *J. Chem. Phys.* 1975, 62:2913.

Knobler, C. M. In *Chemical Thermodynamics*, vol. 2, ed. M. L. McGlashan, ch. 7. London: Burlington House, 1978.

———— and R. L. Scott. *J. Chem. Phys.* 1978, 68:2017.

Kobe, K. A., and J. J. McKetta, ed. *Advances in Petroleum Chemistry and Refining*, vols. 1–4. New York: Interscience, 1958–61.

Kohler, F. *The Liquid State*. Weinheim: Verlag Chemie, 1972.

————. *Monatsh. Chem.* 1957, 88:388, 857.

———— and J. Fischer. *Discuss. Faraday Soc.* 1967, 43:32.

Kohn, J. P., and F. Kurata. *A. I. Ch. E. J.* 1958, 4:211.

Kollman, P. A., and L. C. Allen. *Chem. Rev.* 1972, 72:283.

Koningsveld, R. *Discuss. Faraday Soc.* 1970, 49:144.

Kreglewski, A. *Bull. Acad. Polon. Sci. Ser. Sci. Chim.* 1963, 11:91, 301; 1965, 13:729.

————. In *The Characterization of Chemical Purity*, ed. L. A. K. Staveley, pp. 51–65. London: Butterworth, 1971.

————. *J. Chim. Phys.* 1980, 77:441.

———— and S. S. Chen. *Adv. Chem. Ser.* 1979, 182:197.

———— and ————. *J. Chim. Phys.* 1978, 75:347.

———— and ————. *A Simple Equation of State for Fluids and Mixtures*. Internal Report, Thermodynamics Research Center, College Station, Tex., 1975.

———— and A. Fajans. *Bull. Acad. Polon. Sci. Ser. Sci. Chim.* 1965, 13:441.

———— and K. R. Hall. *Fluid Phase Equil.* 1982, 9:205; 1983, 15:11.

———— and A. Jackowski. *Bull. Acad. Polon. Sci. Ser. Sci. Chim.* 1966, 14:319.

———— and W. B. Kay. *J. Phys. Chem.* 1969, 73:3359.

———— and R. C. Wilhoit. *J. Phys. Chem.* 1974, 78:1961; 1975, 79:449.

————; ————; and B. J. Zwolinski. *J. Phys. Chem.* 1973, 77:2212.

———— and W. Woycicki. *Bull. Acad. Polon. Sci. Ser. Sci. Chim.* 1963, 11:645.

Krichevski, I. R. *Acta Phys-Chim. U.S.S.R.* 1940, 12:480.

———— and D. S. Tsiklis. *Acta Phys-Chim. U.S.S.R.* 1943, 18:264.

Ksiazczak, A., and H. Buchowski. *Fluid Phase Equil.* 1980, 5:131, 141.

Ku, P. S., and B. F. Dodge. *J. Chem. Eng. Data* 1967, 12:158, 168.

Kudchadker, A. P.; G. H. Alani; and B. J. Zwolinski. *Chem. Rev.* 1968, 68:659.

Kuenen, J. D. *Theorie der Verdampfung und Verflussigung von Gemischen.* Leipzig: Barth Verlag, 1906.

Kuenen, J. P., and W. G. Robson. *Phil. Mag.* 1899, 48:180.

Kurata, M., and S. Isida. *J. Chem. Phys.* 1955, 23:1126.

Kwon, O. D.; D. M. Kim; and R. Kobayashi. *J. Chem. Phys.* 1977, 66:4925.

Lacombe, R. H., and I. C. Sanchez. *J. Phys. Chem.* 1976, 80:2352, 2368.

Lal, M., and D. Spencer. *J. Chem. Soc. Faraday Trans. II* 1973, 69:1502.

Lambert, M., and M. Simon. *Physica* 1962, 28:1191.

Lan, S. S., and G. A. Mansoori. *Statistical Thermodynamic Approach to the Prediction of Vapor-Liquid Equilibria of Multicomponent Systems.* Chicago: Univ. of Illinois Dept. of Energy Eng., 1975.

Leach, J. W. Ph.D. diss., Rice Univ., 1967.

————; D. S. Chappelear; and T. W. Leland, Jr. *Proc. Am. Pet. Inst. (Div. Refining)* 1966, 46:223.

Lebowitz, J. L. *Phys. Rev.* 1964, A-133:895.

———— and J. S. Rowlinson. *J. Chem. Phys.* 1964, 41:133.

Lee, J. I., and A. E. Mather. *Can. J. Chem. Eng.* 1972, 50:95.

Lee, J. K.; D. Henderson; and J. A. Barker. *Mol. Phys.* 1975, 29:429.

Lee, M. W.; P. Neufeld; and J. Bigeleisen. *J. Chem. Phys.* 1977, 67:5639.

Lee, S. M.; P. T. Eubank; and K. R. Hall. *Fluid Phase Equil.* 1978, 1:219.

Le Fevre, R. J. W. *Dipole Moments.* London: Methuen, 1948.

Leland, T. W., Jr., and P. S. Chappelear. In *Applied Thermodynamics.* Washington, D.C.: Amer. Chem. Soc., 1968.

————; J. S. Rowlinson; and G. A. Sather. *Trans. Faraday Soc.* 1968, 64:1447.

Lennard-Jones, J. E., and A. F. Devonshire. *Proc. Roy. Soc. (London)* 1937, A-163:53; 1938, A-165:1.

Leonard, P. J.; D. Henderson; and J. A. Barker. *Trans. Faraday Soc.* 1970, 66:2439.

Letcher, T. M. In *Chemical Thermodynamics*, vol. 2, ed. M. L. McGlashan, ch. 2. London: Burlington House, 1978.

Leung, S. S., and R. B. Griffiths. *Phys. Rev.* 1973, 8:2670.

Levelt-Sengers, J. M. H. In *Experimental Thermodynamics*, vol. 2 (IUPAC Monograph), ch. 2. London: Butterworth, 1975.

———. *Ind. Eng. Chem. F.* 1970, 9:470.

Levine, B. F., and C. G. Bethea. *J. Chem. Phys.* 1976, 65:2429, 2439.

Lewis, G. N., and M. Randall. *Thermodynamics*. New York: McGraw-Hill, 1923.

Li, I. P. C.; B. C. Y. Lu; and E. C. Chen. *J. Chem. Eng. Data* 1973, 18:305.

Liang, K.; H. Eyring; and R. Marchi. *Proc. Nat. Acad. Sci. U.S.A.* 1964, 52:1107.

Liebermann, E., and V. Fried. *Ind. Eng. Chem. F.* 1972, 11:350, 354.

Lin, S. H. In *Physical Chemistry*, vol. 5, ed. H. Eyring, ch. 8. New York: Academic Press, 1970.

Lin, Y. N.; R. J. J. Chen; P. S. Chappelear; and R. Kobayashi. *J. Chem. Eng. Data* 1977, 22:402.

Linder, B. *Adv. Chem. Phys.* 1967, 12:225.

———. *J. Chem. Phys.* 1960, 33:668; 1962, 37:963; 1964, 40:2003.

Lippert, J. L.; M. W. Hanna; and P. J. Trotter. *J. Am. Chem. Soc.* 1969, 91:4035.

Locke, D. C. *Adv. Chromatog.* 1976, 14:87.

London, F. *Z. Phys.* 1930, 63:245; *Z. Physik. Chem. (B)* 11:222.

Longuet-Higgins, H. C. *Discuss. Faraday Soc.* 1953, 15:73.

———. *Proc. Roy. Soc.* 1951, A-205:247.

——— and B. Widom. *Mol. Phys.* 1964, 8:549.

Lorimer, J. W.; B. C. Smith; and G. H. Smith. *J. Chem. Soc. Faraday Trans. I* 1975, 71:2232.

Ma, S. M., and H. Eyring. *J. Chem. Phys.* 1965, 42:1920.

McClellan, A. L. *Tables of Experimental Dipole Moments*. London: Freeman, 1963; Vol. 2, Rahara Enterprises, 1974.

McClure, D. W.; K. L. Lewis; R. C. Miller; and L. A. K. Staveley. *J. Chem. Ther.* 1976, 8:785.

McDowell, C. A. In *Physical Chemistry*, vol. 3, ed. Eyring et al., ch. 9. New York: Academic Press, 1969.

McGlashan, M. L. In *Experimental Thermochemistry*, vol. 2, ed. H. A. Skinner, ch. 15. New York: Interscience, 1962.

———. *Trans. Faraday Soc.* 1970, 66:18.

———. *The Use and Misuse of the Laws of Thermodynamics*. Exeter, U.K.: University of Exeter, 1965.

——— and K. W. Morcom. *Trans. Faraday Soc.* 1961, 57:581.

———; ———; and A. G. Williamson. *Trans. Faraday Soc.* 1961, 57:601.

——— and D. J. B. Potter. *Proc. Roy. Soc. (London)* 1962, A-267:478.

———; J. E. Prue; and I. E. Sainsbury. *Trans. Faraday Soc.* 1954, 50:1284.

————; K. Stead; and C. Warr. *J. Chem. Soc. Faraday Trans. II* 1977, 73:1889.

———— and A. G. Williamson. *Trans. Faraday Soc.* 1961, 57:588.

McMicking, J. H., and W. B. Kay. API Division of Refining Conference, Paper no. 18-65, presented at Montreal, Canada, 1965.

Maczynski, A.; Z. Maczynska; and coworkers, *Verified Vapor-Liquid Equilibrium Data.* Warsaw: Polish Sci. Publ., 1976–82.

Magiera, B., and W. Brostow. *J. Phys. Chem.* 1971, 75:4041.

Maher, P. J., and B. D. Smith. *J. Chem. Eng. Data* 1979, 24:17.

Malanowski, S. *Fluid Phase Equil.* 1982, 8:197; 9:311.

Malesinski, W. *Azeotropy.* London: Interscience, 1965.

Mandel, F. *J. Chem. Phys.* 1973, 59:3907.

Mansoori, G.A. *J. Chem. Phys.* 1972, 56:5335.

———— and I. Ali. *Chem. Eng. J. (Netherlands)* 1974, 7:173.

————; N. F. Carnahan; K. E. Starling; and T. W. Leland, Jr. *J. Chem. Phys.* 1971, 54:1523.

———— and T. W. Leland, Jr. *Trans. Faraday Soc.* 1972, 68:320.

Margenau, H., and N. R. Kestner. *Theory of Intermolecular Forces.* London: Pergamon Press, 1969.

Marsh, K. N. In *Chemical Thermodynamics*, vol. 2, ed. M. L. McGlashan, ch. 1. London: Burlington House, 1978.

————. *J. Chem. Ther.* 1971, 3:355; 1977, 9:719.

————; M. L. McGlashan; and C. Warr. *Trans. Faraday Soc.* 1970, 66:2453.

Mason, J. G.; S. N. Naldrett; and O. Maass. *Can. J. Res.* 1940, 18-B:103.

Massengill, D. R., and R. C. Miller. *J. Chem. Ther.* 1973, 5:207.

Massih, A. R., and G. A. Mansoori. *Fluid Phase Equil.* 1983, 10:57.

Mathews, J. F. *Chem. Rev.* 1972, 72:71.

Mattingley, B. I.; Y. P. Handa; and D. V. Fenby. *J. Chem. Ther.* 1975, 7:169, 307.

Mayer, J. E. *J. Chem. Phys.* 1938, 6:87, 101.

———— and M. G. Mayer. *Statistical Mechanics.* New York: Wiley, 1940.

Metropolis, N.; A. W. Rosenbluth; M. N. Rosenbluth; A. H. Teller; and E. Teller. *J. Chem. Phys.* 1953, 21:1087.

Meyer, E. F., and F. A. Baiocchi. *J. Chem. Ther.* 1978, 10:823.

———— and C. A. Hotz. *J. Chem. Eng. Data* 1976, 21:274.

————; T. A. Renner; and K. S. Stec. *J. Phys. Chem.* 1971, 75:642.

———— and R. E. Wagner. *J. Phys. Chem.* 1966, 70:3162.

Michels, A.; M. Geldermans; and S. R. de Groot. *Physica* 1946, 12:105.

Mie, G. *Ann. Physik.* 1903, 11:657.

Miller, R. C. *J. Chem. Phys.* 1971, 55:1613.

————; A. J. Kidnay; and M. J. Hiza. *J. Chem. Ther.* 1977, 9:167.

———— and L. A. K. Staveley. *Adv. Cryog. Eng.* 1976, 21:493.

Minkin, V. I.; O. A. Osipov; and Y. A. Zdanov. *Dipole Moments in Organic Chemistry*. New York: Plenum Press, 1970.

Moelvyn-Hughes, E. A. *Physical Chemistry*. London: Pergamon Press, 1957.

Moldover, M. R., and J. S. Gallagher. In *Phase Equilibria and Fluid Properties in the Chemical Industry*, ch. 30. ACS Symp. Ser. 60. Washington, D.C., 1977.

Mollerup, J. *Fluid Phase Equil.* 1981, 7:121.

——. *J. Chem. Soc. Faraday Trans. I* 1975, 71:2351.

—— and J. S. Rowlinson. *Chem. Eng. Sci.* 1974, 29:1373.

Monfort, J. P., and M. de L. Rojas. *Fluid Phase Equil.* 1978, 2:181.

Morcom, K. W., and D. N. Travers. *Trans. Faraday Soc.* 1965, 61:230.

Morris, J. W.; P. J. Mulvey; M. M. Abbott; and H. C. van Ness. *J. Chem. Eng. Data* 1975, 20: 403, 406.

Morrison, G. *J. Phys. Chem.* 1981, 85:759

—— and C. M. Knobler. *J. Chem. Phys.* 1976, 65:5507.

Moser, B.; K. Lucas; and K. E. Gubbins. *Fluid Phase Equil.* 1981, 7:153.

Mousa, A. H. N.; W. B. Kay; and A. Kreglewski. *J. Chem. Ther.* 1972, 4:301.

Mruzik, M. R.; F. F. Abraham; and G. M. Pound. *J. Chem. Phys.* 1978, 69:3462.

Mulliken, R. S. *J. Chem. Phys.* 1935, 3:573.

—— and W. B. Person. *Molecular Complexes*. New York: Wiley, 1969. In *Physical Chemistry*, vol. 3, ed. D. Henderson, ch. 10. New York: Academic Press, 1969.

Münster, A. *Classical Thermodynamics*. London: Wiley-Interscience, 1970.

Murray, R. S., and M. L. Martin. *J. Chem. Ther.* 1972, 4:723; 1978, 10:613, 711.

Nagata, I.; H. Asano; and K. Fujiwara. *Fluid Phase Equil.* 1978, 1:211.

—— and J. Koyabu. *Thermoc. Acta* 1981, 48:187.

——; K. Tamura; and S. Tokuriki. *Thermoc. Acta* 1981, 47:315.

Nakata, M.; T. Dobashi; N. Kuwahara; and M. Kaneko. *J. Chem. Soc. Faraday Trans. II* 1982, 78:1801.

Naldrett, S. N., and O. Maass. *Can. J. Chem.* 1940, 18-B:118.

Neau, E., and A. Peneloux. *Fluid Phase Equil.* 1981, 6:1; 1982, 8:251.

Neff, R. O., and D. A. McQuarrie. *J. Phys. Chem.* 1973, 77:413; 1975, 79:1022.

Ng, H. J., and D. B. Robinson. *J. Chem. Eng. Data* 1978, 23:325.

Nigam, R. K.; B. S. Mahl; and P. P. Singh. *J. Chem. Ther.* 1972, 4:41.

Nitta, T.; E. A. Turck; R. A. Greenkorn; and K. C. Chao. In *Phase Equilibria and Fluid Properties in the Chemical Industry*. ACS Symp. Ser. 60. Washington, D.C., 1977.

Ochi, K., and B. C. Y. Lu. *Fluid Phase Equil.* 1978, 1:185.

Ohgaki, K., and T. Katayama. *J. Chem. Eng. Data* 1976, 21:53.

Orwoll, R. O., and P. J. Flory. *J. Am. Chem. Soc.* 1967, 89:6814.

Ott, J. B.; J. R. Goates; J. Reeder; and R. B. Shirts. *J. Chem. Soc. Trans. I* 1974, 70:1325.

Parsonage, N. G., and R. L. Scott. *J. Chem. Phys.* 1962, 37:304.

Patterson, D.; Y. B. Tewari; and H. P. Schreiber. *J. Chem. Soc. Faraday Trans. I* 1972, 68:885.

Pauling, L. *The Nature of the Chemical Bond*. Ithaca, N.Y.: Cornell Univ. Press, 1967.

Pavlicek, J., and T. Boublik. *Fluid Phase Equil.* 1981, 7:1, 15.

———; I. Nezbeda; and T. Boublik. *Czech. J. Phys.* 1981, B29:1061.

Peng, D. Y., and D. B. Robinson. *Ind. Eng. Chem. F.* 1976, 15:59.

Percus, J. K., and G. L. Yevick. *Phys. Rev.* 1958, 110:1.

Philippe, R., and P. Clechet. *Third International Conference on Chemical Thermodynamics*, vol. 3, p. 118. Baden, Austria, 1973.

Pierotti, G. J.; C. H. Deal; and E. L. Derr. *Ind. Eng. Chem.* 1959, 51:95.

Pierotti, R. A. *J. Phys. Chem.* 1963, 67:1840.

Pimentel, G. C., and A. L. McClellan. *The Hydrogen Bond*. San Francisco: Freeman, 1960.

Pitzer, K. S. *Adv. Chem. Phys.* 1959, 2:59.

———. *J. Chem. Phys.* 1939, 7:583.

———; D. Z. Lippmann; R. F. Curl, Jr.; C. M. Huggins; and D. E. Petersen. *J. Am. Chem. Soc.* 1955, 77:3433.

Poettmann, H., and D. L. Katz. *Ind. Eng. Chem.* 1945, 37:847.

Pompe, A., and T. H. Spurling. *Virial Coefficients of Mixtures of Hydrocarbons*. Melbourne, Australia: C.S.I.R.O., 1976.

Ponce, L., and H. Renon. *J. Chem. Phys.* 1976, 64:638.

Pool, R. A. H.; G. Saville; T. M. Herrington; B. D. C. Shields; and L. A. K. Staveley. *Trans. Faraday Soc.* 1962, 58:1692.

Pope, G. A.; P. S. Chappelear; and R. Kobayashi. *J. Chem. Phys.* 1973, 59:423.

Pople, J. A. *Proc. Roy. Soc.* 1954, A-221:498, 508.

Powell, R. J.; F. L. Swinton; and C. L. Young. *J. Chem. Ther.* 1970, 2:105.

Pratt, L. R., and O. Chandler. *J. Chem. Phys.* 1977, 67:3683.

———; C. S. Hsu; and D. Chandler. *J. Chem. Phys.* 1978, 68:4202.

Prausnitz, J. M. *Molecular Thermodynamics of Fluid-Phase Equilibria*. Englewood Cliffs, N.J.: Prentice-Hall, 1969.

———. NBS Technical Note 316. Washington, D.C., 1965.

——— and P. L. Chueh. *Computer Calculations of High-Pressure Vapor-Liquid Equilibria*. New York: Prentice-Hall, 1968.

——— and A. L. Myers. *A. I. Ch. E. J.* 1963, 9:5.

Prigogine, I.; A. Bellemans; and V. Mathot. *The Molecular Theory of Solutions*. Amsterdam: North-Holland, 1957.

———— and R. Defay. *Chemical Thermodynamics*. London: Longmans, Green, 1954.

Rall, W., and K. Schaffer. *Ber. Bunsenges. Phys. Chem.* 1959, 63:1019.

Reamer, H. H.; B. H. Sage; and W. N. Lacey. *Ind. Eng. Chem.* 1951, 43:976.

Redlich, O.; E. L. Derr; and G. J. Pierotti. J. Am. Chem. Soc. 1959, 81:2283.

———— and A. T. Kister. *Ind. Eng. Chem.* 1948, 40:381.

———— and J. N. S. Kwong. *Chem. Rev.* 1949, 44:233.

Reid, R. C., and T. K. Sherwood. *The Properties of Gases and Liquids*. New York: McGraw-Hill, 1958.

Reif, F. *Fundamentals of Statistical and Thermal Physics*. New York: McGraw-Hill, 1965.

Reis, J. C. R. *J. Chem. Soc. Faraday Trans. II* 1982, 78:1595.

Reiss, H. *Adv. Chem. Phys.* 1967, 9:1.

————; H. L. Frisch; and J. L. Lebowitz. *J. Chem. Phys.* 1959, 31:369.

Renon, H., and J. M. Prausnitz. *A. I. Ch. E. J.* 1968, 14:135.

Riccardi, R., and M. Sanesi. *Gaz. Chim. Ital.* 1966, 96:542.

Ricci, J. E. *The Phase Rule and Heterogeneous Equilibrium*. New York: Dover, 1966.

Rice, O. K. *J. Phys. & Colloid Chem.* 1950, 54:1293.

Ritchie, G. L. D., and J. Vrbancich. *J. Chem. Soc. Faraday Trans. II* 1980, 76:648, 1245.

Robinson, D. B.; A. P. Lorenzo; and C. A. Macrygeorgos. *Can. J. Chem. Eng.* 1959, 37:212.

Rock, H., and W. Schroeder. *Z. Physik. Chem. (Frankfurt)* 1957, 11:41.

Rodriguez, A. T., and D. Patterson. *J. Chem. Soc. Faraday Trans. II* 1982, 78:501.

Rogalski, M., and S. Malanowski. *Fluid Phase Equil.* 1977, 1:137; 1980, 5:97.

Rogers, B. L., and J. M. Prausnitz. *Trans. Faraday Soc.* 1971, 67:3474.

Rowlinson, J. S. *Chem. Eng.*, Nov. 1974, p. 718.

————. In *Handbuch der Physik*, vol. 12. Berlin: Springer Verlag, 1958.

————. *Liquids and Liquid Mixtures*, 2nd ed. London: Butterworth, 1959, 1969.

————. *The Perfect Gas*. New York: Macmillan, 1963.

———— and M. J. Richardson. *Adv. Chem. Phys.* 1959, 2:85.

———— and J. R. Sutton. *Proc. Roy. Soc.* 1955, A-229:271.

———— and I. D. Watson. *Chem. Eng. Sci.* 1969, 24:1565.

Rushbrooke, G. S. *J. Chem. Phys.* 1963, 39:842.

————; G. Stell; and J. S. Hoye. *Mol. Phys.* 1973, 26:1199.

Salem, L. *J. Chem. Phys.* 1962, 37:2100.

Salsburg, Z. W., and W. Fickett. Los Alamos Sci. Lab. Report LA-2667. Washington, D.C.: U.S. Dept. of Commerce, 1962.

————; P. J. Wojtowicz; and J. G. Kirkwood. *J. Chem. Phys.* 1957, 26:1533.

Sanchez, I. C., and R. H. Lacombe. *J. Phys. Chem.* 1976, 80:2352.

Sanyal, N. K.; D. Ahmad; and L. Dixit. *J. Phys. Chem.* 1973, 77:2552.

Sarkies, K. W.; P. Richmond; and B. W. Ninham. *Aust. J. Phys.* 1972, 25:367.

Saville, G. *J. Chem. Soc. Faraday Trans. II* 1977, 73:1122.

Scatchard, G. *Trans. Faraday Soc.* 1937, 33:160.

Schafer, K. *Ber. Bunsenges. Phys. Chem.* 1977, 81:891.

Schmidt, T. W., and J. P. Guillory. *J. Chem. Phys.* 1976, 64:2027.

Schneider, G. M. *Adv. Chem. Phys.* 1970, 1:17.

————. In *Chemical Thermodynamics*, vol. 2, ed. M. McGlashan, ch. 4. London: Burlington House, 1978.

Scott, G. D. *Nature* 1960, 188:908; 1962, 194:956.

Scott, R. L. *Ber. Bunsenges. Phys. Chem.* 1972, 76:296.

———— In *Chemical Thermodynamics*, vol. 2, ed. M. L. McGlashan, ch. 8. London, Burlington House, 1978.

————. *J. Chem. Phys.* 1956, 25:193.

————. *J. Chem. Soc. Faraday Trans. II* 1977, 73:356.

————. *J. Phys. Chem.* 1971, 75:3843.

———— and D. V. Fenby. *Ann. Rev. Phys. Chem.* 1969, 20:111.

———— and P. H. van Konynenburg. *Discuss. Faraday Soc.* 1970, 49:87.

Sengers, J. V., and A. Levelt-Sengers. *Chem. Eng. News,* June 10, 1968, p. 104.

Sevcik, M.; T. Boublik; and J. Biros. *Collection Czech. Chem. Commun.* 1978, 43:2242.

Shana'a, M. Y., and F. B. Canfield. *Trans. Faraday Soc.* 1968, 64:2281.

———— and P. P. Singh. *J. Chem. Eng. Data* 1975, 20:360.

Sherwood, A. E., and J. M. Prausnitz. *J. Chem. Phys.* 1964, 41:413, 429.

Simnick, J. J.; H. M. Lin; and K. C. Chao. *Adv. Chem. Ser.* 1979, 182:209.

Sinanoglu, O. *Adv. Chem. Phys.* 1967, 12:283.

Singh, A., and A. S. Teja. *Chem. Eng. Sci.* 1976, 31:404.

Singh, B. D., and S. K. Sinha. *J. Chem. Phys.* 1977, 67:3645; 1978, 69:2709, 2927; 1979, 70:552.

Slater, J. C., and J. G. Kirkwood. *Phys. Rev.* 1931, 37:682.

Smith, B. D. *Design of Equilibrium Stage Processes.* New York: McGraw-Hill, 1963.

Smith, J. W. *Electrical Dipole Moments.* London: Butterworth, 1955.

Smith, K. M.; A. M. Rulis; G. Scoles; and R. A. Aziz. *J. Chem. Phys.* 1977, 67:155.

Smoluchowski, M. *Ann. Phys.* 1908, 25:205.

Snider, N. S. *J. Chem. Phys.* 1966, 45:378.

————— and T. M. Herrington. *J. Chem. Phys.* 1967, 47:2248.

Solen, K. A.; P. L. Chueh; and J. M. Prausnitz. *Ind. Eng. Chem. P. D. D.* 1970, 9:310.

Spear, R. P.; R. L. Robinson; and K. C. Chao. *Ind. Eng. Chem. F.* 1969, 8:2.

Specovius, J.; M. A. Leiva; R. L. Scott; and C. M. Knobler. *J. Phys. Chem.* 1981, 85:2313.

Speedy, R. J. *J. Chem. Soc. Faraday Trans. II* 1977, 73:714.

Sprow, F. B., and J. M. Prausnitz. *A. I. Ch. E. J.* 1966, 12:780.

Spurling, T. H., and E. A. Mason. *J. Chem. Phys.* 1967, 46:322.

Stankiewicz, D. W., and A. Kreglewski. *Bull. Acad. Polon. Sci. Ser. Sci. Chim.* 1963, 11:465.

Stanley, H. E. *Introduction to Phase Transitions and Critical Phenomena.* New York: Oxford Univ. Press, 1971.

Starling, K. E. *Fluid Thermodynamic Properties for Light Petroleum Systems.* Houston, Gulf, 1973.

Staveley, L. A. K. *J. Chem. Phys.* 1970, 53:3136.

—————; W. I. Tupman; and K. R. Hart. *Trans. Faraday Soc.* 1955, 51:323.

Stecki, J., and F. Gradzki. *Bull. Acad. Polon. Sci. Ser. Sci. Chim.* 1971, 19:395, 403.

————— and A. W. Jackowski. *J. Chem. Ther.* 1976, 8:1095.

Stein, A., and G. F. Allen. *J. Phys. Chem. Ref. Data* 1973, 2:443.

Stell, G. *Mol. Phys.* 1969, 16:209.

—————; J. C. Rasaiah; and H. Narang. *Mol. Phys.* 1972, 23:393; 1974, 27:1393.

Stephenson, J. In *Physical Chemistry*, vol. 8-B, ed. D. Henderson, ch. 10. New York: Academic Press, 1971.

Stogryn, D. E., and A. P. Stogryn. *Mol. Phys.* 1966, 11:371.

Streett, W. B. *J. Chem. Phys.* 1965, 42:500.

Stryjek, R.; P. S. Chappelear; and R. Kobayashi. *J. Chem. Eng. Data* 1974, 19:340.

Swietoslawski, W. *Azeotropy and Polyazeotropy.* London: Pergamon Press, 1963.

—————. *Ebulliometric Measurements.* New York: Reinhold, 1945.

————— and A. Kreglewski. *Bull. Acad. Polon. Sci. Ser. Sci. Chim.* 1954, 2:77, 187.

Swinton, F. L. In *Chemical Thermodynamics*, vol. 2, ed. M. L. McGlashan, ch. 5. London: Burlington House, 1978.

Tancrede, A. S.; P. Bothorel; P. de St. Romain; and D. Patterson. *J. Chem. Soc. Faraday Trans. II* 1977, 73:15.

Tancrede, P.; D. Patterson; and V. Lam. *J. Chem. Soc. Faraday Trans. II* 1975, 71:985.

Teja, A. S. *Chem. Eng. Sci.* 1975, 30:435.

—— and P. Rice. *J. Chem. Eng. Data* 1976, 21:173.

—— and J. S. Rowlinson. *Chem. Eng. Sci.* 1973, 28:529.

Thiele, E. *J. Chem. Phys.* 1963, 39:474.

Thoen, J.; E. Bloemen; and W. van Dael. *J. Chem. Phys.* 1978, 68:735.

Thompson, P. A. *Equations of State in Engineering and Research*, Adv. Chem. Ser. 1979, 182:365. Washington, D.C.: ACS.

Thomsen, E. S. *Acta Chem. Scand.* 1971, 25:260, 265.

Thorp, N., and R. L. Scott. *J. Phys. Chem.* 1956, 60:670, 1441.

Tiffin, D. L.; A. L. DeVera; K. D. Luks; and J. P. Kohn. *J. Chem. Eng. Data* 1978, 23:45.

Timmermans, J. *The Physico-Chemical Constants of Binary Systems*, vols. 1–4. New York: Interscience, 1959.

Todheide, K., and E. U. Franck. *Z. Physik. Chem.* (Frankfurt), 1963, 37:387.

Tompa, H. *Polymer Solutions*. London: Butterworth, 1956.

Tsiklis, D. S. *Dokl. Akad. Nauk. U.S.S.R.* 1952, 86:993, 1159.

—— and L. A. Rott. *Russ. Chem. Rev.* 1967, 35:351.

Tsimering, L., and A. S. Kertes. *Thermoc. Acta* 1975, 12:206.

Twu, C. H., and K. E. Gubbins. *Chem. Eng. Sci.* 1978, 33:879.

——; ——; and C. G. Gray. *J. Chem. Phys.* 1976, 64:5186.

——; ——; and ——. *Mol. Phys.* 1975, 29:713.

van der Waals, J. D. *Arch. Neerl. Sci. Exactes Natur.* 1890, p. 24; *Z. Physik. Chem.* 1890, 5:133.

——. *Zit. Kon. Acad. Wetensch. (Amsterdam)* 1894, p. 133.

—— and P. Kohnstamm. *Lehrbuch der Thermodynamik*. Leipzig, 1908.

van der Waals, J. H., and J. C. Platteeuw. *Advances in Chemical Physics*, vol. 2, ed. I. Prigogine, p. 1. New York: Interscience, 1959.

van Dranen, J. *J. Chem. Phys.* 1952, 20:1175.

van Laar, J. J. *Z. Physik. Chem.* 1910, 72:723.

van Ness, H. C. *Classical Thermodynamics of Nonelectrolyte Solutions*. New York: Wiley, 1967.

——. In *Reduction of Vapor-Liquid Equilibrium Data*, p. 11. Marseilles, France: U.E.R. Scientifique de Luminy, 1976.

——; S. M. Beyer; and R. E. Gibbs. *A. I. Ch. E. J.* 1973, 19:238.

Vera, H. H., and J. M. Prausnitz. *J. Chem. Eng. Data* 1971, 16:149.

Verlet, L., and J. J. Weis. *Phys. Rev.* 1972, A-5:939.

Vincentini-Missoni, M. In *Phase Transitions and Critical Phenomena*, ed. C. Domb and M. S. Green, ch. 2. New York: Academic Press, 1972.

Wagner, W. *Cryog.* 1972, 12:214.

Wall, L. A.; J. H. Flynn; and S. Straus. *J. Phys. Chem.* 1970, 74:3241.

Walters, G. K., and W. M. Fairbank. *Phys. Rev.* 1956, 103:262.

Wang, S. C. *Physik. Z.* 1927, 28:663.

Wang, S. S.; P. A. Egelstaff; C. G. Gray; and K. E. Gubbins. *Chem. Phys. Lett.* 1974, 24:453.

Warowny, W., and J. Stecki. *The Second Cross Virial Coefficients of Gaseous Mixtures*. Warsaw: Polish Sci. Publ., 1979.

Watson, I. D., and J. S. Rowlinson. *Chem. Eng. Sci.* 1969, 24:1575.

Watts, R. O., and D. Henderson. *Mol. Phys.* 1969, 16:217.

Weeks, J. D. *J. Chem. Phys.* 1977, 67:3106.

————; D. Chandler; and H. C. Andersen. *J. Chem. Phys.* 1971, 54:5237.

Weinberger, M. A., and W. G. Schneider. *Can. J. Chem.* 1952, 30:422.

Weir, R. D.; J. Wynn-Jones; J. S. Rowlinson; and G. Saville. *Trans. Faraday Soc.* 1967, 63:1320.

Wertheim, M. S. *Bull. Am. Phys. Soc.* 1963, 8:329.

————. *J. Math. Phys.* 1964, 5:643.

Wichterle, I. *Fluid Phase Equil.* 1977, 1:161; 1978, 1:225.

———— and R. Kobayashi. *J. Chem. Eng. Data* 1972, 17:4, 9, 13.

————; Z. W. Salsburg; P. S. Chappelear; and R. Kobayashi. *Chem. Eng. Sci.* 1971, 26:1141.

Widom, B. *J. Chem. Phys.* 1965, 43:3898.

Wieczorek, S., and J. Stecki. *J. Chem. Ther.* 1978, 10:177.

Wilcock, R. J.; R. Battino; W. F. Danforth; and E. Wilhelm. *J. Chem. Ther.* 1978, 10:817.

Wilhelm, E. *J. Chem. Phys.* 1973, 58:3558; 1974, 60:3896.

———— and R. Battino. *Chem. Revs.* 1973, 73:1.

Williamson, A. G., and R. L. Scott. *J. Phys. Chem.* 1961, 65:275; 1962, 66:621.

Wilson, G. M. *J. Am. Chem. Soc.* 1964, 86:127.

————. In *Phase Equilibria and Fluid Properties in the Chemical Industry*, ch. 22. ACS Symp. Ser. 60. Washington, D.C., 1977.

Wojtowicz, P. J.; Z. W. Salsburg; and J. G. Kirkwood. *J. Chem. Phys.* 1957, 27:505.

Woycicki, W. *J. Chem. Ther.* 1975, 7:77, 1007.

————. *Proc. Intern. Conf. on Calorimetry and Thermodynamics*, p. 797. Warsaw: PWN, 1969.

Yao, J.; R. A. Greenkorn; and K. C. Chao. *J. Chem. Phys.* 1982, 76:4657.

Young, C. L. In *Chemical Thermodynamics*, vol. 2, ed. M. L. McGlashan, ch. 3. London: Burlington House, 1978.

————. *J. Chem. Soc. Faraday Trans. II* 1972, 68:580.

Zawidzki, J. Z. *Physik. Chem.* 1900, 35:129.

Zawisza, A. *Bull. Acad. Polon. Sci. Ser. Sci. Chim.* 1967, 15:291, 299.

———— and S. Glowka. *Bull. Acad. Polon. Sci. Ser. Sci. Chim.* 1970, 18:549, 555; 1971, 19:191.

Zwanzig, R. W. *J. Chem. Phys.* 1955, 23:1915.

Zwolinski, B. J., et al. *Selected Values of Properties of Hydrocarbons and Related Compounds*, API-44 and TRC Research Projects, Thermodynamics Research Center, Texas A&M University, College Station, Texas, 1961–. (Earlier tables of $Z(T,P)$ and other residual properties have been completely revised since 1968.)

Index